GEOLOGY Made Simple

William H. Matthews

Advisory editor
Eric A. Jarman, B.Sc.

Made Simple Books
W. H. ALLEN   London
A division of Howard & Wyndham Ltd

© 1967 by Doubleday & Company, Inc., and completely reset
and revised 1970 by W. H. Allen & Company Ltd.

Made and printed in Great Britain
by Butler & Tanner Ltd., Frome and London
for the publishers W. H. Allen & Company Ltd.,
44 Hill Street, London, W1X 8LB

First edition, September, 1970
Revised reprint, November, 1974

ISBN 0 491 01552 6 Paperbound

Foreword

The mass media of television and the press bring before the public spectacular news of earthquakes and volcanoes, of discoveries of natural gas, oil, mineral deposits or prehistoric dinosaurs. The study of such events is all part of geology, a subject which is becoming increasingly important as man recognizes the need to explore to the utmost the resources of the planet on which he lives.

In *Geology Made Simple* the general reader will find a clear, ordered approach which will enable him to work alone, either for pleasure or for a specific purpose. The student of a formal course, such as G.C.E. 'O' or 'A' level, will find the book invaluable as background reading and reference; the university undergraduate, beginning his first year of geology, will find it a comprehensive introduction to the subject.

The text is written in two parts: *physical geology* and *historical geology*. Physical geology gives a detailed study of rock-forming minerals and the rocks that are formed by them; it also discusses the forces which act on and within them. Historical geology examines the fossil content of the rock and reconstructs the earth's history over the past 4,500 million years. The appendixes contain much practical advice to the working geologist; they also contain a table of the physical properties of minerals, a glossary of geological terms, and suggestions for further reading.

Geology is an outdoor subject and throughout the book the reader is encouraged to go out and make discoveries and observations for himself, supplementing his theoretical knowledge by practical field-work.

ERIC JARMAN

Table of Contents

CHAPTER THREE
IGNEOUS ROCKS AND VOLCANISM

CHAPTER FOUR
SEDIMENTARY ROCKS

CHAPTER FIVE
METAMORPHISM AND CRUSTAL DEFORMATION

CHAPTER SIX

WEATHERING AND SOIL FORMATION

CHAPTER SEVEN

GEOLOGIC AGENTS: WATER

CHAPTER EIGHT

GEOLOGIC AGENTS: GLACIERS, WIND AND GRAVITY

CHAPTER NINE

OCEANS AND SHORELINES

PART I
PHYSICAL GEOLOGY

CHAPTER ONE

THIS EARTH OF OURS

We live on a wonderful planet called the earth, yet how little most of us really know about its composition and history. We consume the products of soils which have been formed by weathered rock ; we use coal, natural gas and oil formed from the remains of pre-historic plants and animals; we enjoy the beauty of precious stones. But these are only a fraction of the valuable materials with which the earth provides us.

Think of the importance of earth products in the development of modern industry. Our vast mineral resources such as lead, iron, coal and petroleum are derived from the earth, and these products have been made more readily available through the application of basic geology and geological engineering.

The earth has also provided us with areas of exceptional beauty. In Britain we have such phenomena as Cheddar Gorge, the Giant's Causeway, Snowdonia, the Lakes . . . then there are the hot springs in New Zealand, the Great Barrier Reef off Australia, and spectacular volcanoes and waterfalls. All of these, and many more, are the results of geologic processes that are still at work within and on the earth today. They are the same processes which began to shape the earth soon after its birth some four or five thousand million years ago.

(1) The Nature and Scope of Geology

What is geology?

Derived from the Greek *geo*, 'earth', plus *logos*, 'discourse', geology is the science which deals with the origin, structure and history of the earth and its inhabitants as recorded in the rocks.

The events of July 1969 took geology out of its original sphere and plunged it into the Space Age: man had set foot on the moon and brought back samples of moon rock for the geologist, the earth scientist, to study. Who knows to what such studies will lead? Shall we have a clearer picture of the origin of both moon and earth? Shall we find rich mineral deposits, or minerals as yet unknown to us?

Perhaps man will have visited other planets within our solar system before the end of the century; this, too, will create fascinating new branches of geology.

But let us return to our own planet. To the geologist, the earth is not simply the globe upon which we live; it is an ever-present challenge to learn more about such things as earthquakes, volcanoes, glaciers, and the meaning of fossils. How old is the earth? Where did it come from? Of what is it made? To answer these questions, the earth scientist must study the evidence of events that occurred millions of years ago. He must then relate his findings to the results of similar events that are happening today. He tries, for example, to determine the location and extent of ancient oceans and mountain ranges, and to trace the evolution of life as recorded in rocks of different ages. He studies the composition of the rocks and minerals forming the earth's crust in an attempt to locate and exploit the valuable economic products that are to be found there.

In pursuing his study of the earth, the geologist relies heavily upon other basic sciences. For example, **astronomy** (the study of the nature and movements of planets, stars and other heavenly bodies) tells us where the earth fits into the universe, and has also suggested several theories as to the origin of our planet. **Chemistry** (the study of the composition of substances and the changes which they undergo) is used to analyse and study the rocks and minerals of the earth's crust. The science of **physics** (the study of matter and motion) helps to explain the various physical forces affecting our earth, and the reaction of earth materials to these forces.

To understand the nature of prehistoric plants and animals we must turn to **biology**, the study of all living forms. **Zoology** provides us with information about the animals, and **botany** gives us some insight into the nature of ancient plants. By using these sciences, as well as others, the geologist is better able to cope with the many complex problems that are inherent in the study of the earth and its history.

The scope of geology is so broad that it has been split into two major divisions: **physical geology** and **historical geology**. For convenience in study, each of these divisions has been subdivided into a number of more specialized branches or subsciences.

The term **earth science** is commonly used in conjunction with the study of the earth. Although earth science includes the study of geology, it also encompasses the sciences of **meteorology** (the study of the atmosphere), **oceanography** (the study of the oceans), and **astronomy**.

(2) The Geology Around Us

How can we learn more about this fascinating earth and the stories to be read from its rocks? Actually it is very simple, for geology is all around us. The geologist's laboratory is the earth beneath him, and each walk through the fields or drive through the country brings us in contact with the processes and materials of geology.

For example, pick up a piece of common limestone. There are probably fossils in it. And these fossils may well represent the remains of animals that lived in some prehistoric sea which once covered the area.

Or maybe you are walking along a river bank. Notice the silt left on the bank after the last high-water stage. This reminds us of the ability of running water to deposit *sediments*—sediments that may later be transformed into rocks. Notice, too, how swift currents have scoured the river banks. The soil has been removed by erosion, the geologic process which is so important in the shaping of the earth's surface features.

Perhaps you see a field of black fertile soil supporting a fine crop of corn. It may surprise you to learn that this dark rich soil may have been derived from an underlying chalky white limestone—still another reminder of the importance of earth materials in our everyday life.

Within recent years, there has been a notable increase of interest in earth science. More people have become aware of the importance of geology in their everyday lives. More people are visiting museums, studying geology to ordinary and advanced levels in schools, reading geology at colleges or universities. More people are borrowing books from libraries, perhaps to put a name to the shell they found in a cliff or quarry, or to learn more about their own evolution. Others take evening classes, and increasing numbers need to know some geology as part of their occupation. Such a knowledge, whether just a passing acquaintance or an intense study, can make the earth we live on a fascinating subject for study.

(3) Studying Geology

Reading books, listening to lectures or watching films will not make one a geologist. They are aids, the records of other people's knowledge and ideas. To become a geologist it is essential for the student to do some geology at first hand. For example, the sections of this book dealing with mineralogy will be interesting only to those who have seen, handled and marvelled at some actual minerals, even if they are only a selection from a school collection or a display at a museum. Geology is essentially the observation and interpretation of events that have occurred and are occurring in our earth.

Much of the early work which established geology as a science was carried out by enthusiastic amateurs. It may be argued that geology lacks the immediate appeal of the other related sciences, where live plants and animals are observed, but no one can deny that the study of the rocks and their contents takes the student into the finest scenery that the world provides. You, the reader, will enjoy your geology more if you combine your reading with excursions into the countryside.

Appendix B at the end of the book offers suggestions as to how to go about this practical work, but for a start all that is needed is a hammer, chisel, notebook, rucksack, a good pair of eyes and adequate clothing for protection against inclement weather. The soils and their vegetation may indicate the type of underlying rock; buildings and walls made of local stone may well be worth observing. Quarries, cliffs, excavations, river beds and banks will yield much useful information if time is taken to observe them closely.

When you have found a few pieces of rock, had the thrill of finding your first fossil (or the frustration of not finding a fossil), you will naturally want to know how and when the rock was formed, what the fossil was, or why there weren't any in that area. When you have seen folded and crumpled rocks, such as those in the West Country, or read of (or even experienced) earthquakes and the like, automatically you will want to know what hidden forces have produced them. This is where reading will help. Like the ancient Greeks who found marine fossils high up in Egyptian mountains, you may well be able to work out for yourself that once upon a time the region was a sea swarming with life. When you are out on the Yorkshire moors and find a piece of coral in a limestone, as you shiver and shelter from the wind, you may cast your mind back to the days when the region was covered by a shallow warm sea and think of today's tropical regions. It is in such circumstances that a book such as this can help you to crystallize your ideas and bring before you other information which you have not been able to observe for yourself.

The amateur geologist enjoys himself thoroughly; but whereas the amateur may be content to spend a couple of hours in a quarry, the professional or exceptionally keen amateur will go over the face inch by inch collecting fossils, pieces of rock, or minerals from different horizons, carefully labelling them and later sorting them equally carefully to see if there is any variation in the population, or the size of grain, of the particles in the rock. He will also extend his observations laterally to see if the rock becomes sandier or coarser, for this may indicate that at the time of formation the region was near a coast-line. He will notice the attitude of the rocks, the direction and angle at which they dip, and by comparing his observations with the attitudes existing elsewhere he may well be able to work out what has happened below the ground.

To find out more accurately what has happened below the ground, a bore-hole may be made and samples of rock brought up from hundreds of feet below the surface. Wells and mines also yield valuable information of the geology beneath the outcropping rocks. Many geologists work for oil companies, mining companies and water boards. From continued observation, and trial and error, the most favourable conditions for the location of oil or natural gas are known. And so the modern prospector goes into action. The pattern of the rock as it is traced across country may suggest that the formation below ground is worth investigating. Alternatively, explosives may be used to produce shock waves like those of a natural earthquake (see Chapter Eleven) so that a picture of the rocks below can be built up as the waves rebound off resistant beds. After this initial survey the drill and rig must be brought into action, which is a costly and time-consuming procedure. The geologist then studies the microfossils that are brought up by the drill, and these too may indicate whether it is worth while continuing the exploratory drilling.

Recent developments with satellites scanning a country or region sometimes indicate unusual formations that have been missed from ground-level exploration. The use of infra-red and other types of cameras, in conjunction with satellites, is also opening up new methods of locating economically important minerals.

But not all the work of the geologist is in the field; much of his time is spent in the laboratory, measuring very accurately the size of specimens, looking for minute changes, or cutting very thin rock slices for observation under the microscope.

(4) Physical Geology

Physical geology deals with the earth's composition, its structure, the movements within and upon the earth's crust, and the geologic processes by which the earth's surface is, or has been, changed.

This broad division of geology, which includes such basic geologic subsciences as **mineralogy** (the study of minerals) and **petrology** (the study of rocks), provides us with much-needed information about the composition of the earth. In addition, it embraces **structural geology** to explain the arrangement of the rocks within the earth, and **geomorphology** to explain the origin of its surface features. Another important branch of physical geology is **economic geology**, the study of the economic products of the earth's crust and their application for commercial and industrial purposes. Included here, for example, are the important fields of mining and petroleum geology. (These and several other branches of physical geology will be considered in the first part of this book.)

The branches of physical geology enable the geologist to make detailed studies of all phases of earth science. The knowledge gained from such research brings about a better understanding of the physical nature of the earth.

(5) Historical Geology

Historical geology is the study of the origin and evolution of the earth and its inhabitants. Like physical geology, historical geology covers such a variety of fields that it has been subdivided into several branches. Each of these branches is actually a science in its own right, and one may devote a lifetime of study to specializing in any one of them.

In working out the geologic history of an area, the geologist utilizes **stratigraphy**, which is concerned with the origin, composition, proper sequence, and correlation of the rock strata. **Palaeontology** (the study of ancient organisms as revealed by their fossils) provides a background to the development of life on earth, and **palaeo-geography** affords a means of studying geographic conditions of past times. It is thus possible to reconstruct the relations of ancient lands and seas and the organisms that inhabited them.

The major branches of historical geology, like those of physical geology, overlap in some areas and are interdependent. The physical geologist uses mineralogy and petrology to determine what types of rocks are present and from what they were derived. The historical geologist studies the same rocks to ascertain what kinds of animals or plants were living at the time the rocks were deposited, what environment they lived in, and what kind of climate then prevailed. Thus, the unification of physical and historical geology leads ultimately to a better understanding of the composition and history of our earth.

(6) The Earth in Space

The earth, its relation to the stars and planets, and speculations as to its origin, attracted man's curiosity long before the birth of the geological sciences. Although such studies fall more properly within the realm of astronomy, a brief survey of the earth and its planetary relations will provide the reader with some understanding of its place in the universe.

Galaxies, disc-shaped clusters containing millions or billions of stars, are the major constituents of the universe. Astronomers estimate that there are large numbers of galaxies in outer space. However, only one, the Milky Way, the galaxy in which the earth is located, need be discussed here. This galaxy, which contains every star that we can see with the naked eye, is lens-shaped and contains thousands of millions of stars, including our sun.

A great mass of gaseous material like the other stars, the sun has a diameter of about 865,000 miles and is located about halfway between the centre and the edge of the Milky Way. The sun is important to man, for it is the centre of the solar system. The solar system is composed of the sun, nine planets (all of which revolve around the sun), thousands of asteroids or planetoids (small planets), comets, and meteors.

The **planets**, the largest of the solid bodies in the solar system, all move on essentially the same plane around the sun. Nine in number, they are (in order from the sun): Mercury, Venus, Earth, Mars, Jupiter, Saturn, Uranus, Neptune and Pluto.

Associated with most of the planets are smaller bodies called **satellites** or **moons**, which revolve around each planet. The moon, earth's satellite, revolves around our planet approximately once every month. Some planets (Mercury, Venus and Pluto) have no known satellites. However, Jupiter, the largest planet, has twelve.

Located between the orbits of Mars and Jupiter are thousands of small planet-like bodies called **asteroids** (or planetoids). They revolve about the sun in much the same manner as the planets.

Meteors, rocklike objects travelling through space, become highly heated when they enter the earth's atmosphere. These so-called 'shooting stars' are usually completely burnt out before they reach the earth's surface. Some, however, strike the earth as **meteorites**.

The solar system also contains large, self-luminous, celestial bodies called **comets**. When comets come near the sun they can be seen from the earth. Unfortunately, because of their eccentric (off-centre) orbits, comets are infrequently visible; but their return to the vicinity of earth is mathematically predicted. For example, probably the best known comet, **Halley's comet**, whose last 'visits' to earth were in 1835 and 1910, is expected to return in 1986.

(7) Shape, Size and Motions of the Earth

The earth is the largest of the four planets of the inner group (Mercury, Venus, Earth and Mars) of the solar system, and is third closest to the sun (Fig. 1).

Shape of the Earth. The earth has the form of an 'oblate spheroid', in other words it is almost ball-shaped, or spherical, except for a

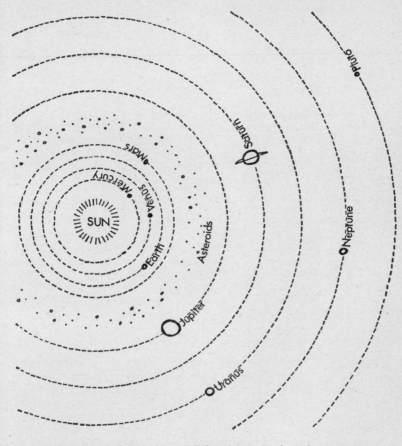

Fig. 1. Relative positions of the planets in the solar system

slight flattening at the poles. This flattening, and an accompanying bulge at the equator, are produced by the centrifugal force of rotation.

Size of the Earth. Earth has a polar diameter of about 7,900 miles (12,650 kilometres). The equatorial diameter is approximately 27 miles (43 kilometres) greater because of the equatorial bulge. Even so, the earth's equatorial diameter is less than that of Jupiter,

Saturn, Uranus and Neptune. The circumference of the earth is about 24,874 miles (39,800 km), and the surface area comprises roughly 197 million square miles (500 million km²) of which only about 51 million square miles (130 million km²), i.e. 29 per cent, is land. The remaining 71 per cent of the earth's surface is covered by water.

Earth Motions. Each of the planets revolves around the sun within its own orbit and with a certain period of revolution. In addition to its trip around the sun, the earth also rotates.

Rotation of the Earth. The earth turns on its **axis** (the shortest diameter connecting the poles), and this turning motion is called rotation. The earth rotates from west to east and makes one complete rotation each day. It is this rotating motion that gives us the alternating periods of daylight and darkness which we know as day and night.

The axis about which the earth rotates is not perpendicular to its orbit around the sun (if it were we should have no changing seasons of the year). In fact, today the axis is inclined at an angle of 23½ degrees to the perpendicular. We say that it is 23½ degrees 'today' because the angle is slowly changing all the time as the earth's axis 'wobbles' backwards and forwards—taking about 26,000 years to complete a single 'wobble'.

Revolution of the Earth. The earth revolves around the sun in a slightly elliptical orbit approximately once every 365¼ days. During this time (a solar year), the earth travels at a speed of more than 60,000 miles (96,000 km) per hour, and on the average it remains about 93 million miles (149 million km) from the sun.

In addition to rotation, revolution and the 'wobbling' motion, the earth is travelling with the entire solar system in the general direction of the star Vega at a speed of about 400 million miles (640 million km) per year.

Principal Divisions of the Earth

The earth consists of air, water and land. We recognize these more technically as the **atmosphere**, a gaseous envelope surrounding the earth; the **hydrosphere**, the waters filling the depressions and covering almost three-quarters of the surface; and the **lithosphere**, the solid part of the earth which underlies the atmosphere and hydrosphere.

The Atmosphere. The atmosphere, or gaseous portion of the earth, extends upward for hundreds of miles above sea level. It is a mixture of nitrogen, oxygen, carbon dioxide, water vapour, and other gases (see Table 1).

The atmosphere makes life possible on our planet. Moreover, it acts as an insulating agent protecting us from the heat and ultra-violet radiation of the sun and shielding us from the bombardment of meteorites, and it makes possible the evaporation and precipitation of moisture. The atmosphere is an important geologic agent (see Chapter Six) and is responsible for the processes of weathering which are continually at work on the earth's surface.

TABLE 1. *Analysis of gases present in pure dry air. Notice that*
nitrogen and oxygen comprise 99 per cent of the total.

GAS	PERCENTAGE BY VOLUME
Nitrogen	78·084
Oxygen	20·946
Argon	0·934
Carbon dioxide	0·033
Neon	0·001818
Helium	0·000524
Methane	0·0002
Krypton	0·000114
Hydrogen	0·00005
Nitrous oxide	0·00005
Xenon	0·0000087

The Hydrosphere. The hydrosphere includes all the waters of the
oceans, lakes and rivers on the surface of the earth, as well as **ground**
water which exists in the pores and crevices of the crustal rocks and
soil. Most of this water is contained in the oceans, which cover
roughly 71 per cent of the earth's surface to an average depth of
about $2\frac{1}{2}$ miles (4 kms).

The waters of the earth are essential to man's existence and they
are also of considerable geologic importance. Running streams and
oceans are actively engaged in eroding, transporting and depositing sediments; and water, working in conjunction with atmospheric agents, has been the major force in forming the earth's
surface features throughout geologic time. The geologic work of the
hydrosphere will be discussed in some detail in later chapters of this
book.

The Lithosphere. Of prime importance to the geologist is the
lithosphere. This, the solid portion of the earth, is composed of rocks
and minerals which, in turn, comprise the continental masses and
ocean basins (see Chapter Nine). The rocks of the lithosphere are of
three basic types: **igneous, sedimentary** and **metamorphic.** Igneous
rocks were originally in a molten state, but have since cooled and
solidified to form rocks such as granite and basalt. Sedimentary
rocks were formed from sediments (fragments of pre-existing rocks)
deposited by wind, water or ice. Limestone, sandstone and clay
are typical of this group. The metamorphic rocks have been transformed from rocks that were either sedimentary or igneous in origin.
This transformation takes place as the rock is subjected to great
physical and chemical change. Marble, which in its original form
was limestone, is an example of a metamorphic rock.

Most of what we know about the lithosphere has been learnt
through the study of the surface materials of the earth. However,
by means of deep bore holes and seismological studies, geologists
have gathered much valuable information about the earth's interior.
Additional geologic data are derived from rocks which were originally buried many miles beneath the ground but have been brought

to or near the surface by violent earth movements and later exposed by erosion.

The lithosphere has been divided into three distinct zones, each of which is described in Chapter Eleven.

(9) Major Physical Features of the Earth

The major relief features of the earth are the **continental masses** and the **ocean basins**. These are the portions of the earth which have apparently remained stable throughout all known geologic time.

The Continental Masses. The continents are rocky platforms which cover approximately 29 per cent of the earth's surface. Composed largely of granite, they have an average elevation of about three miles (five kilometres) above the floors of the surrounding ocean basins and rise an average of half a mile (one km.) above sea level

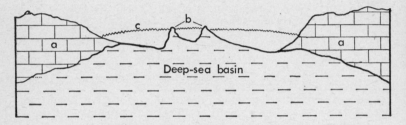

Fig. 2. Continents and ocean basins: (*a*) continents, (*b*) volcanic islands, (*c*) sea level

(Fig. 2). The seaward edges of the continental masses are submerged, and these are called the **continental shelves.**

Although the continental surfaces appear very irregular to man, the difference in elevation between the highest mountain (Mount Everest is just over 29,000 feet above sea level) and the deepest part of the ocean (more than 35,000 feet below sea level, south of the Mariana Islands in the Pacific) is inconsequential when considered in relation to the size of the earth. The land forms responsible for the earth's surface irregularities are discussed in Chapter Twelve.

The Ocean Basins. The ocean basins contain the greatest part of the hydrosphere and cover more than 70 per cent of the earth's surface. The floors of the oceans are not, as was once believed, flat and featureless; in fact they possess as many irregularities as the land and include deep trenches, canyons and submarine mountain ranges.

Of the five oceans (Arctic, Southern or Antarctic, Atlantic, Indian and Pacific) the Pacific is deepest (about 35,000 feet) and largest, covering almost half the earth. The bottoms of the deepest parts of the oceans are composed of a rather dense, dark, igneous rock called basalt. In many places the basaltic bottom is covered by layers of marine sediment.

The origin of the continents and ocean basins and their relationship to each other are discussed in later chapters.

(10) Geologic Forces

Geologic investigation of almost any part of the earth's surface will reveal some indication of great changes which the earth has undergone. These changes are of many kinds and most have taken place over millions of years. They are, in general, brought about by the processes of **gradation, tectonism** and **volcanism.**

Gradation. The surface rocks of the earth are constantly being affected by 'gradational' forces. For example, the atmosphere attacks the rocks, weathering them both physically and chemically. In addition, the rivers and oceans of the hydrosphere are continually wearing away rock fragments and transporting them to other areas where they are deposited. Gradation, then, includes two separate types of process: **degradation,** which is a wearing-down or destructive process, and **aggradation,** a building-up or constructive process.

Degradation, commonly referred to as **erosion,** results from the wearing down of the rocks by water, air and ice. Here are included the work of atmospheric weathering, glacial abrasion, stream erosion, wind abrasion and so on.

Aggradation, known also as **deposition,** results in the accumulation of sediments and the ultimate building up of rock strata. The principal agents depositing these sediments are wind, ice and water. The work of each of these geologic agents is discussed elsewhere in this book.

Tectonism. This term encompasses all the movements of the solid parts of the earth with respect to each other. Tectonic movements, which are indicative of crustal instability, produce **faulting** (fracture and displacement), **folding, subsidence** and **uplift** of rock formations. Known also as **diastrophism**, tectonism is responsible for the formation of many of our great mountain ranges and for most of the structural deformation that has occurred in the earth's crust. However, these tectonic features (such as folds and faults) are not usually seen until they have been exposed by the process of erosion.

In addition, widespread tectonic movements are responsible for certain types of metamorphism (see Chapter Five). The intrusion of **magma**, more closely associated with volcanism (see below), may also bring about rock deformation by folding.

Volcanism. This term, known also as **vulcanism,** refers to the movement of molten rock materials within the earth or on the surface of the earth. Volcanic processes produce the lavas, ashes and cinders which are ejected from volcanoes. Volcanism is also responsible for the plutonic rocks, once molten, which solidified at great depth within the earth (see Chapter Three).

CHAPTER TWO

MINERALS

The geologist is primarily interested in the earth's rocky crust; therefore he must know something about **minerals**, for these are the building blocks of the earth's crust. Although geologists differ when defining the term 'mineral', the following definition is generally accepted: *minerals are chemical elements or compounds which occur naturally within the crust of the earth.* They are **inorganic** (not derived from living things) and therefore this definition does not include coal or petroleum. They have a definite chemical composition or range of composition, an orderly internal arrangement of atoms (crystalline structure) and certain other distinct physical properties. It should be noted, however, that the chemical and physical properties of some minerals may vary within definite limits.

Rocks are aggregates or mixtures of minerals, the composition of which may vary greatly. **Limestone,** for example, is composed primarily of one mineral—**calcite. Granite,** on the other hand, typically contains three minerals—**feldspar** (sometimes spelt felspar), **mica** and **quartz.**

Certain minerals, such as calcite, quartz and feldspar, are so commonly found in rocks that they are called the **rock-forming** minerals. Other minerals, like **gold, diamond, uranium** minerals and **silver,** are found in relatively few rocks.

Minerals vary greatly in their chemical composition and physical properties. Let us now examine the more important physical and chemical characteristics that enable us to distinguish one mineral from another.

(1) Chemical Composition of Minerals

Although a detailed discussion of chemistry is not within the scope of this book*, an introduction to chemical terminology is necessary if we are to understand the chemical composition of minerals.

All matter, including minerals, is composed of one or more elements. An element is a substance that cannot be broken down into simpler substances by ordinary chemical means. Theoretically, if you were to take a quantity of any element and cut it into smaller and smaller pieces, eventually you would obtain the smallest pieces that still retained the characteristics of the element. These minute particles are **atoms.** Although atoms are so small that they cannot be seen with the most powerful microscope, we know a great deal

*For a comprehensive introduction to general chemistry, see *Chemistry Made Simple* by Fred C. Hess (W. H. Allen).

14

about them. We know, for instance, that the nucleus of an atom is composed of **protons,** positively charged particles, and **neutrons,** or uncharged particles. Outside the nucleus and revolving rapidly around it are negatively charged particles called **electrons.** It is now known, of course, that certain elements can be broken down by atomic fission or 'atom-smashing', but these are not considered to be 'ordinary chemical means'. Although there are only ninety-two elements occurring in nature, several more have been created artificially.

Some minerals, such as gold or silver, are composed of only one element. More often, however, minerals consist of two or more elements united to form a **compound.** For example, calcite is a chemical compound known as calcium carbonate. The chemical composition of a compound may be expressed by means of a chemical formula ($CaCO_3$ in the case of calcite) in which each element is represented by a symbol. For many elements, the first letter of the element's name is used as its symbol—thus, H for an atom of hydrogen, and C for an atom of carbon. If the names of two elements start with the same letter, two letters may be used for one of them to distinguish between their symbols. For example, an atom of helium may be represented as He, an atom of calcium as Ca. Some symbols have been derived from an abbreviation of the Latin name of the elements: Cu (from *cuprum*) represents an atom of copper, and Fe (from *ferrum*) an atom of iron. The small numerals used in a chemical formula represent the proportion in which each element is present. Hence, the formula for water, H_2O, indicates that there are two atoms of hydrogen for each atom of oxygen present in water.

Although ninety-two elements have been found to occur in nature, the eight of them named below are so abundant that they constitute more than 98 per cent, by weight, of the earth's crust.

ELEMENT SYMBOL	PERCENTAGE BY WEIGHT
Oxygen (O)	46·60
Silicon (Si)	27·72
Aluminium (Al)	8·13
Iron (Fe)	5·00
Calcium (Ca)	3·63
Sodium (Na)	2·83
Potassium (K)	2·59
Magnesium (Mg)	2·09
Total	98·59

As indicated in the above table, two elements, oxygen and silicon, make up approximately three-quarters of the weight of the rocks. Both these elements are **non-metals,** but the remaining six are **metals.** Metals are characterized by their capacity for conducting heat and electricity, their ability to be hammered into thin sheets (malleability) or to be drawn into wire (ductility), and their lustre (the way light is reflected from the mineral's surface). Such minerals as gold, silver, copper and iron are included in the metals. The non-metallic minerals do not have the properties mentioned above: typical examples are sulphur, diamond and calcite.

(2) Crystals

When minerals solidify and grow without interference, they will normally adopt smooth, angular, symmetrical shapes known as crystals. The planes that form the outside of the crystals are known

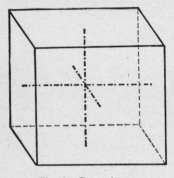

as **faces.** The shape of the crystal and the angles between adjacent sets of crystal faces are related to the arrangement of atoms within the crystal and are important in mineral identification.

Crystal Systems. Each mineral has been assigned to one of six crystal systems. These systems have been established on the basis of the number, position, and relative lengths of the **crystal axes**—imaginary lines extending through the centre of the crystal (Figs. 3–9). For example, crystals assigned to the 'tetragonal' system (see below) have three axes: two of equal

Fig. 3. Crystal axes

length called the *horizontal* axes; the third axis, which may be longer or shorter than the other two, is called the *vertical* axis because it is always placed vertically when the crystal is properly oriented (see Fig. 5).

Any given crystal system possesses a type of symmetry that is peculiar to all members of that system, but is unlike that of crystals belonging to any other systems. The type of symmetry present is determined by the arrangement of the axes of the crystal.

Mineralogists recognize the following six crystal systems:

(i) *Isometric or Cubic System.* Crystals belonging to this system have three axes of equal length at right-angles to one another (see below).

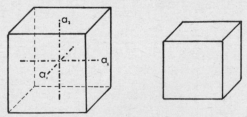

Fig. 4. An isometric crystal (halite); the axes are labelled *a, b, c*

(ii) *Tetragonal System.* Tetragonal crystals also have three axes mutually perpendicular to each other. The two lateral (horizontal) axes are equal in length but the third (vertical) axis is either longer or shorter than these (see below).

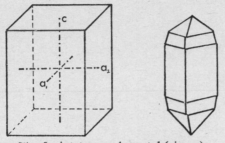

Fig. 5. A tetragonal crystal (zircon)

(iii) *Orthorhombic System.* Crystals assigned to this class have three axes all at right-angles and each of a different length (see below).

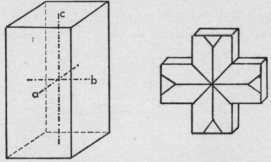

Fig. 6. An orthorhombic crystal (staurolite)

(iv) *Hexagonal System.* This system has crystals typified by three horizontal axes of equal length which intersect at angles of 120°, and a vertical axis at right-angles to these. The vertical axis is either longer or shorter than the three horizontal axes (see below).

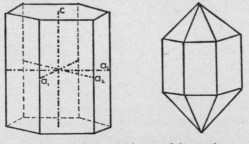

Fig. 7. A hexagonal crystal (quartz)

(v) *Monoclinic System.* Monoclinic crystals have three unequal axes, two of which intersect at right-angles. The third axis is oblique to the plane of the others (see below).

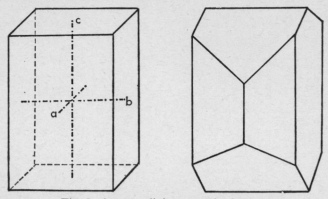

Fig. 8. A monoclinic crystal (orthoclase)

(vi) *Triclinic System.* The triclinic system has crystals characterized by three axes of unequal length and all oblique to one another (see below).

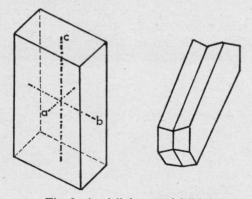

Fig. 9. A triclinic crystal (albite)

Crystallography. The study of crystals is a specialized subject in its own right. It has been proved mathematically that it is possible to have only 32 types of symmetry in crystals. The common minerals fall into 11 groups which can be placed in the six crystal systems outlined above. Some of the 32 varieties of symmetry are only known in artificial compounds or have no known representative as yet. The faces of the crystals in the accompanying diagrams are numbered according to an index system devised by Miller (another system is due to Weiss).

For example, in Fig. 10 the face 101 cuts the a_1 axis in a positive sense and the c axis in a negative sense, but does not cut the a_2 axis. It should be noted that the Miller symbols are based on the reciprocal distances: thus a face which is parallel to an axis cuts it at infinity and has Miller symbol 0. Another example (Fig. 11) shows a common monoclinic crystal with only the forward faces numbered.

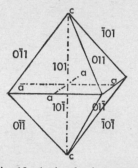
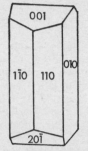

Fig. 10. A simple zircon crystal (tetragonal system)

Fig. 11. Orthoclase (monoclinic system)

Obviously numbering the faces is quite a complicated procedure, and students who need to know more about it will find it explained in books devoted to mineralogy.

Crystal Habits. Any given mineral crystal will grow in such a way as to assume a characteristic shape or form. This typical form is called the habit of the crystal. For example, galena has a **cubic** habit (Fig. 12a), tourmaline a **columnar** habit (Fig. 12b), and barytes (barite) a **tabular** habit (Fig. 12c). Because crystals of a specific

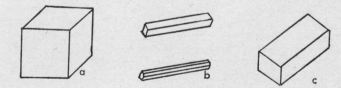

Fig. 12. Crystal habits: (a) cubic, (b) columnar, (c) tabular

mineral will develop only in the crystal system of that mineral, it follows that a cubic-system crystal will show only cubic-system characteristics. However, when crystals are formed at different temperatures, they may assume different habits within their system. Thus, fluorite crystals formed at low temperatures have a cubic form, while those formed at high temperatures have an octahedral form. In some instances a mineral crystal may show a combination of two forms.

B

In addition to temperature, such factors as pressure, the composition of the solutions from which the minerals crystallize, and variation in mineral composition, may affect crystal forms. The presence of impurities in a mineral may also modify the habit of a crystal.

(3) Physical Properties of Minerals

Each mineral possesses certain physical properties or characteristics by which it may be recognized or identified. Although some may be identified by visual examination, others must be subjected to certain simple tests.

Physical properties especially useful in mineral identification are (a) hardness, (b) colour, (c) streak, (d) lustre, (e) specific gravity, (f) cleavage, (g) fracture, (h) shape or form, (i) tenacity or elasticity, and (j) taste, odour and feel. There are also certain other properties which will be discussed. The geologist must know how to test a mineral specimen if he is to identify it correctly. Many of these tests do not require expensive laboratory equipment and may be carried out in the field. Some of them may be made by using such commonplace articles as a knife or a hardened-steel file, a copper coin, a small magnet, an inexpensive pocket lens with a magnification of six to ten times, a piece of glass, a piece of unglazed porcelain tile—even a fingernail.

Hardness. One of the easiest ways to distinguish one mineral from another is by testing for hardness. The hardness of a mineral is determined by what materials it will scratch, and what materials will scratch it. The hardness or scratch test may be done with simple testing materials carried in the field. For greater accuracy, one may use the **Mohs scale** of hardness, devised by the nineteenth-century German mineralogist, Friedrich Mohs. In studying his mineral collection, Mohs noticed that certain minerals were much harder than others. He believed that this variation could be of some value in mineral identification, so he selected ten common minerals to be used as standards in testing other minerals for hardness. In establishing this scale, Mohs assigned each of the reference minerals a number. He designated talc, the softest in the series, as having a hardness of 1. The hardest mineral, diamond, was assigned a hardness of 10.

The Mohs scale, composed of the ten reference minerals arranged in order of increasing hardness, is as follows:

1—Talc (softest)
2—Gypsum
3—Calcite
4—Fluorite
5—Apatite
6—Feldspar
7—Quartz
8—Topaz
9—Corundum
10—Diamond (hardest)

Most of the minerals in Mohs' scale are common ones, which can be obtained cheaply and easily. Diamond chips are more expensive, but not beyond reason. Note that the Mohs scale is so arranged that each mineral will be scratched by those having higher numbers, and will scratch those having lower numbers.

It is also possible to test for hardness by using the following common objects:

OBJECT	HARDNESS
Finger-nail	about $2\frac{1}{2}$
Copper coin	about 3
Glass	$5-5\frac{1}{2}$
Knife blade	$5\frac{1}{2}-6$
Steel file	$6\frac{1}{2}-7$

Each of the above will scratch a mineral of the indicated hardness. For example: the finger-nail will scratch talc (hardness of 1) and gypsum (hardness of 2), but would not scratch calcite which has a hardness of 3.

In testing for hardness, first use the more common materials. Start with the finger-nail; if that will not scratch the specimen, use the knife blade. If the knife blade produces a scratch, this indicates that the specimen has a hardness of between $2\frac{1}{2}$ and 6 (see scale above). By referring to the Mohs scale, it will be found that there are three minerals of known hardness within this range. These are, apatite (5); fluorite (4); and calcite (3). If the calcite will not scratch the specimen but the fluorite will, its hardness is further limited to between 3 and 4. Next, try to scratch the fluorite with the specimen. If this can be done, even with difficulty, the hardness is established as 4; if not, then it is between 3 and 4.

Colour. Probably one of the first things that is noticed about a mineral is its colour. However, the same mineral may vary greatly in colour from one specimen to another, and, with certain exceptions, colour is of limited use in mineral identification. Certain minerals have relatively constant colours, for example, azurite, which is always blue; malachite, which is green; and pyrite, which is yellow. Other minerals, such as quartz or tourmaline, occur in a wide variety of colours; hence, colour may be of little use in identifying these two minerals. Colour variations of this sort are generally due to the presence or absence of chemical impurities in the mineral.

When using colour in mineral identification, it is necessary to take into consideration such factors as (i) whether the specimen is being examined in natural or artificial light, (ii) whether the surface being examined is fresh or weathered, and (iii) whether the mineral is wet or dry. Each of these may cause colour variations in a mineral. In addition, certain of the metallic minerals will tarnish and the true colour will not be revealed except on a fresh surface.

Streak. When a mineral is rubbed across a piece of unglazed tile, it may leave a line similar to a pencil or crayon mark. This line is composed of the powdered mineral. The colour of this powdered material is known as the 'streak' of the mineral, and the unglazed tile used in such a test is called a **streak plate** (Fig. 13).

The streak of some minerals will not be the same as the colour of the specimen. For example, a piece of black haematite (sometimes spelt hematite) will leave a reddish-brown streak. An extremely hard

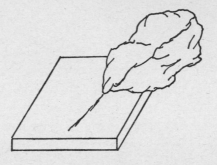

Fig. 13. Testing for streak

mineral such as topaz or corundum will leave no streak at all, since the streak plate (hardness about 7) is softer than either topaz (8) or corundum (9): thus it will be the plate that is powdered, not the mineral.

Lustre. The appearance of the surface of a mineral as seen in reflected light is called lustre. A few minerals, such as native silver and gold, are said to have metallic lustre. Other lustres are called non-metallic. The more important non-metallic lustres and some common examples are shown below:

Adamantine—brilliant glossy lustre: typical of diamond
Vitreous—glassy, looks like glass: quartz or topaz
Resinous—the lustre of resin: sphalerite
Greasy—like an oily surface: nepheline
Pearly—like mother-of-pearl: talc
Silky—the lustre of silk or rayon: asbestos or satin-spar gypsum
Dull—as the name implies: chalk or clay

Sub-metallic lustre is intermediate between metallic and non-metallic lustre. The mineral wolframite displays typical sub-metallic lustre.

Terms such as **shining** (bright by reflected light), **glistening** (having a sparkling brightness), **splendent** (having a glossy brilliance), and **dull** (lacking brilliance or lustre), are commonly used to indicate the degree of lustre present. Here, too, one must take into consideration such factors as tarnish, type of lighting, and general condition of the mineral specimen being examined.

Specific Gravity. The density or specific gravity of a mineral is a useful clue to its identification. Specific gravity is determined by comparing the weight of the mineral specimen with the weight of an equal volume of fresh water. Thus, a specimen of the lead ore called galena (specific gravity about 7·5) would be about $7\frac{1}{2}$ times as heavy as the same volume of water.

In order to determine its specific gravity, the specimen is first weighed in air on a spring-balance, then it is lowered into a container of fresh water and weighed again. The true weight of the specimen (i.e. its weight in air) is divided by the difference between the two readings to give its specific gravity. Then the result can be compared with a table of known specific gravities of minerals.

Cleavage and Fracture. Mineral crystals will break if they are strained beyond their plastic and elastic limits. If the crystal breaks irregularly it is said to exhibit 'fracture'; but if it should break along surfaces related to the crystal structure it is said to show 'cleavage'.

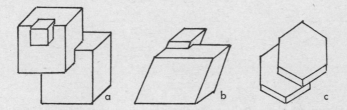

Fig. 14. Types of cleavage: (*a*) cubic, (*b*) rhombic, (*c*) perfect basal

Each break or **cleavage plane** is closely related to the atomic structure of the mineral and designates planes of weakness within the crystal. Because the number of cleavage planes present and the angles between them are constant for any given mineral, cleavage is a very useful aid in mineral identification.

Minerals may have one, two, three, four or six directions of cleavage. The mineral galena, for example, cleaves in three planes (directions) at right-angles to one another. Thus, if galena is struck a quick, sharp blow with a hammer, the specimen will break up into a number of small cubes. Calcite, on the other hand, has three cleavage planes that are not at right-angles to one another: when it breaks, therefore, it will produce a number of rhombohedral cleavage fragments. Hence, galena is said to have *cubic* cleavage, calcite *rhombohedral* cleavage.

Many minerals break or fracture in a distinctive way, and for this reason their broken surfaces (Fig. 15) may be of value in identifying them.

There are several types of fracture; some of the more common types (with examples) are:

Conchoidal—the broken surface of the specimen shows a fracture resembling the smooth curved surface of a shell. This type of fracture is typical of chipped glass, quartz and obsidian.
Splintery or Fibrous—fibres or splinters are revealed along the fracture surface. Examples: pectolite and asbestos.

Hackly—fracture surface marked by rough jagged edges: copper and silver.

Uneven—rough irregular fracture of surface. This type of fracture is common in many minerals and is, therefore, of limited use in identification: jasper (a variety of quartz).

Even—as the name implies: magnesite.

Earthy—as the name implies: kaolinite.

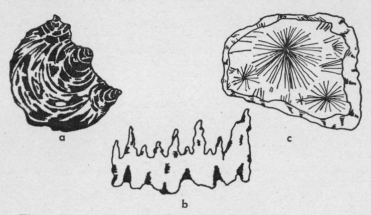

Fig. 15. Types of fracture: (*a*) conchoidal, (*b*) hackly, (*c*) splintery

Tenacity. The tenacity of a mineral may be defined as the resistance that it offers to tearing, crushing, bending, or breaking. Some terms used to describe the different kinds of tenacity are:

Brittle—the mineral can be broken or powdered easily. The degree of brittleness may be qualified by such terms as 'tough' or 'fragile': galena or sulphur.

Elastic—the mineral, after being bent, will return to its original form or position: mica.

Flexible—the mineral will bend but will not return to its original shape when the pressure is removed: talc.

Sectile—the mineral can be cut with a knife to produce shavings: selenite, gypsum and talc.

Malleable—the mineral can be hammered into thin sheets: gold and copper.

Ductile—the mineral can be drawn out into wire: gold, silver and copper.

Taste. Some minerals which are soluble have a distinctive taste. It is not unusual for a student to lick a piece of mineral which he thinks might be halite, commonly called rock-salt, to see if it has a saline taste. Chile saltpetre (sodium nitrate) has a cooling taste, whereas alum, which is readily soluble, has a sweet astringent taste, and Epsom salt (magnesium sulphate) has a bitter taste. Other minerals have their own peculiar flavour.

Odour. When certain minerals are struck, rubbed, breathed upon or heated, a distinct odour may be produced. When pyrites is struck or heated, a strong smell of sulphur is emitted. A 'clayey' or argillaceous smell is given off when clay is breathed upon. Other indicative odours include the smell of garlic produced when an arsenic compound is heated.

Feel. The feel of minerals may help in their identification. Some are smooth or greasy; others are harsh or rough. Certain minerals adhere to the tongue.

Other Physical Properties. In addition to those properties already discussed, the mineral characteristics below may also aid greatly in identification. Examples of minerals exhibiting these properties are given.

Play of Colours. Some minerals show variations in colour when viewed from different angles: labradorite.

Asterism. This may be observed if the mineral exhibits a starlike effect when viewed either by reflected or transmitted light: certain specimens of phlogopite or the star-sapphire.

Transparency. Known also as *diaphaneity*, this property refers to the ability of a mineral to transmit light. The varying degrees of transparency are: (i) **opaque**—no light passes through the mineral (galena, pyrites and magnetite); (ii) **translucent**—light passes through the mineral but an object cannot be seen through it (chalcedony and certain other varieties of quartz); (iii) **transparent**—light passes through the mineral and the outline of objects can be clearly seen through it (halite, calcite and clear crystalline quartz).

Double refraction. When light enters a mineral it may be refracted in two directions producing a double image when looked through (Fig. 19). Calcite is the best known mineral to exhibit this phenomenon.

Magnetism. A mineral is said to be magnetic if, in its natural state, it will be attracted to an iron magnet: magnetite, pyrrhotite. Lodestone, a form of magnetite, is naturally magnetized and is repelled by an iron magnet.

Luminescence. When a mineral glows or emits light that is not the direct result of incandescence, it is said to be luminescent. This phenomenon is usually produced by exposure to ultra-violet rays. Exposure to X-rays, cathode rays, or radiation from radioactive substances can also cause luminescence. If the mineral is luminous only during the period of exposure to the ultra-violet rays or other stimulus, it is said to be **fluorescent** (scheelite and willemite, for example). A mineral which continues to glow after the cause of excitation has been removed is said to be **phosphorescent.**

Twinning. A crystal may be composed of two or more distinct parts oriented in different directions and joined along a 'twin-plane'. If the crystal is composed of two halves it is said to be **simple.** This may be seen in a specimen of Shap granite containing large crystals or orthoclase feldspar. Orthoclase is twinned according to three laws: the *Carlsbad, Baveno* and *Manebach* laws (see Fig. 16). Plagioclase, another feldspar, has a repeated twin in accordance with the *Albite* law. **Penetration twins,** typified by the 'Iron Cross' form of pyrite, are crystals that have to some extent coalesced, with the remaining faces of one parallel to those of its twin.

Twinning is, of course, best studied under the microscope.

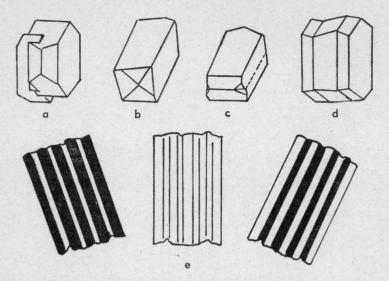

Fig. 16. Twin laws: (*a*) Carlsbad, (*b*) Baveno, (*c*) Manebach, (*d*) Albite, (*e*) Albite twins in plagioclase as seen in three positions under a microscope

(4) The Petrological Microscope

Anyone who takes more than a casual interest in geology will need to learn how to use the petrological microscope. This instrument enables the student to view mineral specimens under ordinary light, under polarized light (light restricted to vibrating in a certain plain, as with polarizing sun-glasses), and between crossed polarizing filters. The use of different types of illumination enables the mineralogist to make a more accurate determination of the mineral content of a rock.

Incidentally, a piece of apparently dull, uninteresting rock, when cut into very thin slices and viewed under polarized light, may well be exceedingly beautiful.

(5) Amorphous Minerals

Although most substances accepted as minerals are crystalline, some lack the ability to crystallize and are commonly said to be amorphous ('without form'). Most amorphous minerals are hard, glassy substances; opal is an example. They are sometimes referred to as **mineraloids.**

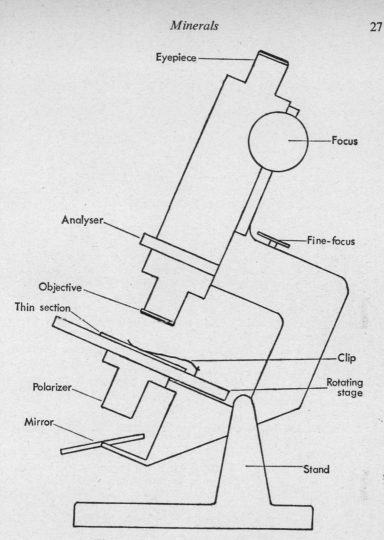

Fig. 17. The petrological microscope

(6) Rock-forming Minerals

Of the 2,000 or so different minerals known to be present in the earth's crust, relatively few are major constituents of the more common rocks. Those minerals that do make up a large part of the more abundant types of rock are called the rock-forming minerals. Most of them are **silicates,** that is, they consist of a metal combined with silicon and oxygen. Some of the most important rock-forming minerals are briefly discussed below: their physical properties are listed in Appendix A on page 250.

(*a*) **Feldspars.** Minerals belonging to the feldspar group constitute the most important group of rock-forming minerals. They are so abundant, in fact, that they have been estimated to make up as much as 60 per cent of the earth's crust. Feldspars are found in almost all igneous rocks as well as in many sedimentary and metamorphic rocks. (The different types of rock are explained in Chapters Three, Four and Five.)

Chemically, the feldspars are silicates of aluminium and one or two other metals. These may be potassium, sodium, calcium, or, more rarely, barium. Orthoclase and plagioclase are the two principal groups of feldspars.

Orthoclase is a rather common potash feldspar. It can usually be distinguished from plagioclase feldspar by absence of striations (stripes). **Microcline** is another potassium aluminium silicate of the same chemical composition ($KAlSi_3O_8$) as orthoclase. However, these two feldspars crystallize in different crystal systems and differ in certain other physical characteristics.

Orthoclase often shows 'simple' twinning under one of the three laws given on page 25. It is frequently found as **phenocrysts** (large crystals) in granite, e.g. Shap granite.

Plagioclase feldspars, known also as soda-lime feldspars, are common in many igneous rocks and in certain metamorphic rocks. Colours range from white, through yellow and reddish-grey, to black. Two varieties of plagioclase, **albite moonstone** and **labradorite**, are characterized by white to bluish internal flashes (a play of colours called *opalescence*).

The feldspars are of considerable commercial importance. Orthoclase is used in the manufacture of china, porcelain and scouring powders. Feldspars are also used in making paints, enamels and glass. Plagioclase feldspars are less commonly used than potash feldspars, but some are employed in the ceramics industry.

(*b*) **Quartz.** Quartz is one of the most widely distributed of all minerals; it forms an important part of many igneous rocks and is common in many sedimentary and metamorphic rocks. Quartz may occur in combination with other minerals, or, as in the case of pure sandstones and quartzites, may be the only one present.

Pure quartz is composed of **silica** (SiO_2), the only oxide of silicon, but certain varieties contain impurities such as iron or manganese. These impurities are responsible for the varied colours of certain types of quartz. Quartz occurs in crystalline aggregates or in irregular grains or masses. The term **crypto-crystalline** (literally 'with hidden crystals') is used to describe quartz varieties in which crystals are not evident.

Crystallizing in the hexagonal system, quartz commonly forms six-sided crystals with pyramidal ends (Fig. 18). Some of the more common crystalline varieties of quartz are amethyst, milky quartz, rose quartz, rock-crystal, cairngorm and smoky quartz. 'Massive' (with no crystalline shape) or crypto-crystalline varieties include agate, chalcedony, chert, flint and jasper.

Quartz tends to occur almost anywhere, and most sands are composed largely of quartz fragments. Quartz crystals are used in certain electronic equipment such as radio oscillators, and some varieties of quartz are used in making glass for lenses and prisms. Other types of quartz are valuable as semi-precious stones or gems. In some cases the variety is peculiar to a district, for example, the manganese stained variety known as Sark-stone, which gives rise to a local industry in the Channel Islands. Sandstone is used as building stone, and quartz sands are used as abrasives and in making concrete and glass.

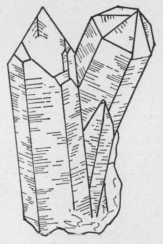

(*c*) **Mica.** Minerals of the mica group are characterized by perfect basal cleavage (commonly called micaceous cleavage). Micas are usually easily identified, as they typically occur in paper-thin, shiny, elastic cleavage plates.

The micas, like the feldspars, are aluminium silicates that are characterized by very complex chemical formulas.

Only two varieties of mica, **muscovite** and **biotite,** are especially important as rock-forming minerals. However, phlogopite and lepidolite are fairly common in some rocks.

Fig. 18. Well developed hexagonal quartz crystals with pyramidal ends

Muscovite, known also as white mica or potash mica, is usually transparent. It occurs typically in thin, elastic, scalelike crystals, and is a common constituent of certain granites and pegmatites (see Chapter Three). It also occurs in certain metamorphic and sedimentary rocks.

India, Russia (whence it gets its name) and America are the biggest producers of muscovite mica. Its shiny, flaky crystals can readily be seen in the more acidic granites such as those of Devon and Cornwall.

Commercially, muscovite is used in manufacturing electrical equipment, insulating cloth and tape, lamp shields, lubricants, paints, and Christmas-tree 'snow'.

Biotite, or black mica, is a very common mica and usually occurs in association with muscovite. It is found in many igneous and metamorphic rocks, where it is seen as thin, platy, shiny black sheets or scales. Biotite is typically dark brown to black (sometimes green), and is a complex silicate of aluminium, calcium, magnesium, and iron. With the exception of the black colour, the physical properties of biotite are essentially the same as those of muscovite. Unlike muscovite, biotite mica has very little commercial value. This was the mineral that the first men on the moon thought they recognized as they looked at their first moon rock.

(*d*) **Pyroxenes.** The pyroxene group is composed of complex silicates, and is among the most common of all rock-forming minerals.

The most widespread pyroxene is **augite,** a common constituent of many of the dark-coloured igneous rocks.

Jadeite found in New Zealand and elsewhere is one of the two varieties of jade, a precious stone.

Pyroxenes are commonly found also in certain metamorphic rocks.

(*e*) **Amphiboles.** The amphiboles, another group of common rock-forming minerals, are closely related to the pyroxenes. Because of their great similarity, these two groups are often confused. Chemically, they are complex silicates containing magnesium, calcium and iron.

Hornblende, the most abundant amphibole, is a common constituent of igneous and metamorphic rocks.

Actinolite and **tremolite** are other interesting amphiboles. Some mineralogists consider these as separate and distinct minerals; others treat them as a single mineral or refer to them as the *tremolite–actinolite series.* They commonly occur in long, prismatic, bladed crystals, or as fibrous or asbestiform (resembling asbestos) masses. Fibrous tremolite is used to some extent as asbestos in fireproofing and insulation. (Tremolite asbestos should not be confused with serpentine or chrysotile asbestos. The latter are more commonly used in industry.)

The other form of jade, **nephrite,** is an amphibole. Also used for ornaments is 'tiger's eye', a variety of another amphibole, **riebeckite,** found in South Africa.

(*f*) **Calcite.** The mineral calcite is composed of calcium carbonate ($CaCO_3$) and is the most common member of the calcite group. It occurs in many sedimentary and metamorphic rocks, and is the primary constituent of most limestones (see Chapter Four). Calcite occurs in crystalline, granular or chalky masses; as a vein mineral; in cave and spring deposits; and in the shells of certain animals (corals, snails and clams for example).

Calcite will **effervesce** or 'fizz' in cold dilute acids; this is a useful test for calcite and some other minerals. Some forms are fluorescent, and certain clear calcite crystals have the property of **double refraction** whereby an object viewed through the crystal appears double (see Fig. 19).

Some of the more common varieties of calcite are Iceland spar, dog-tooth spar, chalk, travertine (including the calcareous tufa deposited by springs, and stalactites and stalagmites formed in caves) and limestone.

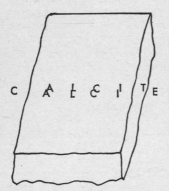

Fig. 19. Rhombohedral calcite crystal exhibiting double refraction

Calcite, as the chief component of limestones and marbles, is used in making cement, lime and plaster, as a flux in the smelting of iron ores, and as building or ornamental stone. It is also used in the manufacture of glass, paint and fertilizer. Certain transparent varieties of calcite are used in making optical instruments—especially as polarizing prisms.

(*g*) **Dolomite.** A compound of calcium carbonate and magnesium carbonate, $CaCO_3.MgCO_3$, dolomite is common in sedimentary rock, where it is often mixed with calcite. It may also occur in association with many ore deposits and in veins or cavities in some igneous rocks. Dolomite differs from calcite in that it is slightly harder ($3\frac{1}{2}$), is only mildly affected by cold dilute acids and may exhibit crystals with curved faces. Its major uses are as a building stone (much 'marble' consists of dolomite), in the manufacture of cement, and as a source of magnesium.

Dolomite is named after a famous French geologist and mineralogist, Guy de Dolomieu (1750–1801). A bed of *magnesium limestone* rich in dolomite runs from Nottingham to the Durham coast.

(*h*) **Fluorite.** Fluorite or fluor-spar consists mainly of calcium fluoride (CaF_2). It occurs in a variety of colours ranging from transparent to almost black, and is widely distributed. In Britain it is mined commercially at Weardale (Durham) and Castleton (Derbyshire). The largest producer of fluor-spar is America. It is used as an enamel, in glass, and in the manufacture of hydrofluoric acid. A bluish-purple variety known as 'bluc-john', and found only at Castleton, is made into vases and jewellery.

(*i*) **Aragonite.** Aragonite, like calcite, is composed of calcium carbonate ($CaCO_3$), but it differs from calcite in that it is less stable and crystallizes in the orthorhombic system. It occurs as a secondary mineral in cavities of limestone; as a deposit around hot springs and geysers; in cave deposits; and in the shells of such animals as clams and corals. Although not as common as calcite, aragonite is used for the same purposes.

(*j*) **Gypsum.** The very common mineral known as gypsum is a hydrated form of calcium sulphate ($CaSO_4.2H_2O$). A product of evaporation, it occurs in thick-bedded deposits in the English Midlands. Among the more common varieties of gypsum are selenite, satin spar, alabaster and rock gypsum.

A mineral of great economic importance, gypsum is extensively used in the manufacture of plaster of Paris and plaster board, and in agricultural fertilizers. It is also used in the manufacture of Portland cement, paint, glass, porcelain and blackboard 'chalks'. The quality of the beers produced in the Burton-on-Trent area is partially attributed to the gypsum content of the water. The alabaster variety of gypsum is used in statuary and as an ornamental stone.

(*k*) **Anhydrite.** Although chemically similar to gypsum, anhydrite ($CaSO_4$) is harder and heavier, and crystallizes in the rhombic form (whereas gypsum is monoclinic). When heated it may show pink, yellow-green or blue-white fluorescence. Commonly found in the cap rock of certain salt domes, anhydrite occurs in extensive beds at Billingham in Durham, where it is mined as a raw material for the chemical industry. It is also used in the manufacture of cement and fertilizer, and to a lesser degree as an ornamental stone.

(*l*) **Halite.** Commonly called rock-salt, halite is composed of sodium chloride (NaCl). Large deposits of halite have been formed as a result of the evaporation of prehistoric inland seas. During the evaporation period, beds of various minerals were laid down in the following sequence: calcite and dolomite formed first; they are covered by gypsum and anhydrite and then, thirdly, rock-salt; these were followed by other, more soluble salts. This sequence has been observed in the Strassfurt deposits in Germany, where it is repeated several times owing to variations in prehistoric climate and conditions. Thus a group of very important minerals are today found deposited in massive beds.

There is one salt mine left in Cheshire (it supplies the rock-salt used for treating icy roads), but more usually water is pumped through the earth forming **brine** which is evaporated to yield common salt. Britain and India are among the half dozen countries that produce over a million tons of salt a year. Apart from its domestic uses it is vital to the chemical industries, glass manufacture, soap and metallurgy. Further, chlorine, which is obtained from salt, is used for purifying water and as a bleaching agent.

Salt differs from other rocks in that instead of fracturing and folding under pressure, it flows. Salt glaciers occur in Persia, and salt domes, which form as a result of salt being squeezed up through weak areas in the earth's crust, provide ideal conditions for the formation of petroleum. This phenomenon is best seen in the coastal regions of Texas and Louisiana.

(*m*) **Kaolin.** Perhaps better known as china-clay, kaolin is a soft, white, rather greasy mineral formed by the decomposition of rock containing feldspar. Its main constituent is a hydrated aluminium silicate called kaolinite ($Al_2O_3.2SiO_2.2H_2O$). Kaolin becomes 'plastic' or mouldable when wet—as it must do, of course, in the hands of a potter. It has the rather unusual property of adhering to the tongue, and emits an earthy, claylike odour when breathed upon.

The Cornish kaolin deposits were probably formed from granite by the pneumatolysis of (action of air on) feldspar. Although kaolin is an aluminium mineral, there is as yet no economical means of extracting the metal from it. Its chief use is in the ceramics (pottery) industry; it is used also in the manufacture of paint, paper and rubber.

(*n*) **Serpentine.** The serpentines are a complex group of hydrated magnesium silicates having the general formula $3MgO.2SiO_2.2H_2O$. They commonly occur in compact masses which feel greasy or soapy. The ordinary form of serpentine is blackish-green in colour, mottled and veined with white and other colours—hence the name. *Verde antico* or 'serpentine marble' is veined with white minerals such as calcite and dolomite. Because it can be highly polished, this stone is much used for small carvings and other ornamental work. A considerable industry has grown up around The Lizard in Cornwall supplying tourists with inexpensive ornaments and costume jewellery. **Chrysotile,** a fibrous variety of serpentine, is the principal mineral used as asbestos.

(*o*) **Chlorite.** The chlorite group is composed of complex silicates of aluminium, magnesium and iron, in combination with water: it has nothing to do with chlorine. These minerals are usually green in colour, and resemble the micas. They generally occur in foliated or scaly masses, although tabular six-sided crystals may be found. They are common constituents of many igneous and metamorphic rocks.

(7) Metallic or Ore Minerals

Metals are among the most valuable products known to man, and this is one reason why the metallic, or ore, minerals are of great interest to the geologist. Metallic minerals are found in ore deposits —rock masses from which metals may be obtained commercially. Usually occurring with the valuable ore minerals are certain worthless minerals called **gangue** minerals. The gangue, of course, must be separated from the more valuable ore.

Some of the more familiar metals and their ores are discussed below.

(*a*) **Aluminium.** One of the most important metals of industry, aluminium is derived primarily from **bauxite.** Bauxite is actually a mixture of two forms of aluminium oxide: diaspore ($Al_2O_3.H_2O$), and gibbsite ($Al_2O_3.3H_2O$). It occurs in earthy, claylike masses, or in a pisolitic (pea-like) form as rounded concretions in a clayey matrix.

Because it is light, resists corrosion, and is relatively strong, aluminium is especially useful in all kinds of application—from the construction of aeroplane frames to the manufacture of household utensils and ornamental objects. It is a good conductor of heat and electricity, hence it is being used to replace the more expensive copper for these purposes. Although aluminium is the most abundant of the metals, comprising 8 per cent of the earth's crust, it is not often in a convenient form. It has already been mentioned that kaolin is a potential source; the alum-shale in the Whitby area of Yorkshire has provided aluminium, but is of little economic value today, as evidenced by the disused quarries.

(*b*) **Copper.** Copper has contributed much to the development of civilization. It is found principally in igneous rocks or in vein deposits and is widely distributed over much of the world. Although there are many different copper minerals (about 165 have been described) only those of major economic importance will be discussed below.

Native copper (Cu) occurs in irregular masses, or plates, in various parts of the world. **Chalcopyrite** ($CuFeS_2$), known also as copper pyrite, is widely distributed in many rocks and is the principal ore of copper. It typically occurs in massive, brass-yellow forms and is found in, among other places, Cornwall and Australia. Chalcopyrite differs from pyrite (iron sulphide) in that it is slightly darker and much softer. The extreme brittleness of chalcopyrite serves to distinguish it from gold, which is always malleable. However, in some localities chalcopyrite may carry gold and also silver.

Chalcocite (Cu_2S) commonly occurs scattered throughout the enclosing rock, and because of this it has been called copper porphyry. It occurs also as vein deposits. Although typically a low-grade ore, chalcocite is usually easily and economically extracted. Chalcocite occurs in commercial quantities in America and elsewhere.

Azurite ($2CuCO_3.Cu(OH)_2$), also known as **chessylite,** is characterized by its azure-blue colour and the tendency to effervesce in acid. A copper carbonate, it usually occurs in smooth or irregular masses and is commonly found associated with malachite (see below). Short, tabular, deep-blue monoclinic crystals may be developed. Like malachite, it is used as an ornamental stone as well as being an important source of copper. It is found in the Redruth area of Cornwall; but it gets the name chessylite from a district in France.

Malachite ($CuCO_3.Cu(OH)_2$), although of similar chemical composition to azurite, is easily distinguished by its bright green colour and its pale-green streak. Malachite is more abundant than azurite and commonly occurs in veins in limestone. It is another important copper ore and has been used to a limited extent in table-tops, vases and other ornamental work. Like azurite, it is found at Redruth in Cornwall, but the largest deposits occur in Siberia, Australia (Burra Burra mine), and Katanga in the Congo Republic.

Other commercially important ores of copper include: **cuprite** (red oxide) mined in Cornwall and Australia; **bornite** (horse-flesh ore) mined in Cornwall and Germany; and **chrysocolla,** also found in Cornwall, Australia (near Adelaide), Katanga and Zambia.

(c) **Gold.** Because of its great beauty and the fact that it is found in the native state and does not have to be extracted from ore by complicated metallurgical processes, gold has been prized by man since the dawn of history. The idea that gold is valuable because of its rarity is a popular misconception. In fact gold is very widely distributed around the world (even in sea water). **Native gold** (Au) is found typically in quartz veins and in association with the mineral pyrite (which it resembles). **Alluvial gold** is found in Cornwall and Scotland, and the gold mined in North Wales supplies the wedding rings for the Royal Family. But British production of gold is insignificant compared with that of South Africa, the U.S.S.R., the United States, Canada or Australia.

(d) **Lead.** The most important source of lead is **galena** (PbS), which occurs in a wide variety of rocks, including igneous, sedimentary and metamorphic. A sulphide of lead, it may be found as a replacement (see page 100) in limestone, in veins, or in localized concentrated 'pockets'. Copper, zinc, and silver ores are often found associated with galena, and some of these ores occur in sufficient amounts to make them valuable commercially. Galena is a soft grey mineral, rather like lead itself.

Most of the world's galena is mined in America and Australia (especially at Broken Hill, New South Wales). It has formed an important industry in the past in Britain, being found most commonly in Cornwall, Cumberland, Derbyshire, the Isle of Man and Scotland.

When found in sufficient quantities, **cerussite** ($PbCO_3$) is another source of lead. A third ore, **anglesite** ($PbSO_4$), is also a valuable source and is found in various parts of Britain including Porys Mine in Anglesey; and, again, in Australia.

Lead is used in the manufacture of paints (in the form of white lead), type metal, pipes, accumulators, solder, metal alloys and as a shielding material to protect against radioactivity and X-rays.

(*e*) **Mercury.** The most abundant ore of mercury (known also as quicksilver) is **cinnabar** (mercuric sulphide, HgS). Although found in relatively few places, cinnabar occurs in both volcanic and sedimentary rocks, and near hot springs. It occurs most typically as red or black fine-granular or earthy masses. Native (or free) mercury may also be found in small silvery droplets in certain cinnabar deposits.

The world's leading producer of mercury is Spain, but Italy is also a relatively large producer. Cinnabar is also produced in America and has been worked in Australia and New Zealand. The fact that cinnabar is deposited from the waters of hot springs, is an indication of their formation in Tertiary volcanic rocks.

Mercury combines with most metals to form 'amalgams': copper amalgam is used for filling teeth. Mercury is important in the manufacture of explosives, and the manufacture of such scientific instruments as thermometers and barometers.

(*f*) **Silver.** Another metal that is highly prized by man, silver may occur as **native silver** (Ag), usually tarnished, either in veins or disseminated throughout the rocks.

Argentite, a silver sulphide (Ag_2S), is the most abundant of silver ores. Found in veins, where it may be associated with free silver and a number of other metallic minerals, it is usually massive or encrusting, but cubic crystals may be formed.

The world's largest producer of silver is Mexico, with the United States second and Canada third. Norway (Kongsberg), Peru, and Australia (Broken Hill, New South Wales, and Mount Isa mines, Queensland) also produce considerable amounts.

Used in making coins, jewellery, and tableware, silver is also used in plating metals, and in the photographic, chemical and electronics industries.

(*g*) **Tin.** The only important tin ore is the oxide **cassiterite** (SnO_2), or **tinstone.** Although widely distributed in small amounts, it occurs in commercial quantities in igneous rocks, where it is commonly associated with quartz, topaz, galena and tourmaline. Tin was mined by the Romans in Cornwall, but nowadays the majority of the world's tin comes from Malaya, Bolivia and Indonesia.

Tin is used for tin plating (the lining of tin cans is one of its major uses), in solder and type metal, tinfoil, and—mixed with copper—for making bronze.

(*h*) **Zinc.** This is another metal of considerable economic importance. The primary ore is **zinc blende** or **sphalerite** (ZnS), a rather common mineral whose origin and occurrence are similar to those of galena, with which it is commonly associated. It is found in veins in igneous, sedimentary, and metamorphic rocks, and as replacement deposits (see page 100) in limestone. The principal producers of zinc are the United States, Canada, Mexico, Peru and Australia. Zinc is used in galvanizing steel, and in the manufacture of paint, brass, cosmetics, type metal, dry cell batteries, and for a multitude of other purposes.

(i) **Iron.** Probably the most essential of all metals, iron is obtained from a variety of minerals including haematite, magnetite, and limonite.

Haematite, ferric oxide (Fe_2O_3), is one of the world's most common minerals; it occurs in massive black beds and in scaly schistose rocks. It is sedimentary in origin, and most deposits have been altered and enriched by subsequent solutions. Haematite has been extracted in various parts of Britain including the Forest of Dean (Gloucestershire) and Cumberland. The largest deposits are in Canada in the region of Lake Superior.

Magnetite (Fe_3O_4) is strongly attracted by a magnet, and a variety of magnetite that acts as a magnet is known as **lodestone.** Large deposits are found in Scandinavia and America.

Limonite is a term used to refer to several commingled hydrous iron oxides (the chemical formula is roughly $Fe_2O_3.H_2O$). It occurs in compact or earthy masses and is a relatively common iron ore. Found in Spain and Cuba, it is also prominent in the British iron-ores of Jurassic age.

Pyrite, or 'fool's gold', is an iron mineral that is little used as a source of iron. It is an iron sulphide (Fe_2S) and is common associated with a number of different ores, including copper and gold. Pyrite is a valuable source of sulphur in the manufacture of sulphuric acid.

Native Iron is found in the Giant's Causeway, in meteorites and elsewhere. **Siderite** ($FeCO_3$) is another important vein mineral worked in Germany. **Marcasite** (FeS_2), found in concretions in chalk, for instance, is used as a semi-precious stone in costume jewellery.

(j) **Nickel.** Nickel is widely used in alloys for coins and in the electrical industry. Like iron, it can be magnetized.

Pentlandite (Fe,Ni)S is the principal ore of nickel: the largest deposits found so far are in Sudbury, Ontario. In December 1969 the Stock Exchange saw the price of shares in the Poseidon mining company rise to unprecedented heights following the discovery of large reserves of nickel in Windorra, Australia.

(k) **Cobalt.** Cobalt ores are often found with those of iron and nickel. **Smaltite** ($CoAs_2$) is mined in Cornwall, Ontario, Katanga and Zambia. Cobalt is used in the chemical industry, and as an important pigment in paints and ceramics. Permanent magnets are nearly always made from cobalt alloys.

(l) **Chromium.** The only important ore of chromium is **chromite** ($FeCr_2O_4$). Among the largest producers of chromite are Russia, Rhodesia and India; it is an important material for lining furnaces. The metal itself is used for chromium plating and in stainless steel.

(m) **Manganese.** Often found with cobalt and its associates, manganese occurs in several oxide ores, two of which are **pyrolusite** and **psilomelane.** The principal sources of these ores are Russia and India. Manganese is of great importance in alloys, especially in the steel industry.

(*n*) **Magnesium.** Widely distributed throughout the world, magnesium is the eighth most common element in the earth's crust. Metallic magnesium is extracted by electrical means from **carnallite** ($KCl.MgCl.6H_2O$) and from sea water.

Magnesite ($MgCO_3$) is found in veins in Greece and India and as a replacement mineral (see page 100) in Austria and Canada. Magnesite is used as a furnace-liner, and in the cement, paper and sugar industries.

Dolomite ($CaCO_3.MgCO_3$), which has been described on page 31, is becoming an economical source of magnesium. Two other ores of magnesium worth noting are **epsomite** (Epsom salt, $MgSO_4.7H_2O$), which is used in medicine, and **spinel** ($MgAl_2O_4$), a gem variety found in Ceylon, Burma and Thailand.

(*o*) **Uranium.** In this so-called 'atomic age', radioactive minerals have come to play an ever-increasing part in modern technology. Although there are a number of radioactive minerals, only two of the more important ones will be considered here.

Uraninite, known also as **pitchblende,** is the most important source of uranium and radium. A complex oxide of uranium, uraninite also contains minor amounts of thorium, lead, helium and certain other relatively rare elements. Uraninite may be found as a primary constituent in certain granites and pegmatites. It is also known to occur as a secondary mineral with ores of lead, copper, and silver.

Carnotite, or potassium uranyl vanadate ($K_2(UO_2)_2(VO_4)_2.3H_2O$) is an earthy, powdery mineral which is an ore of uranium and vanadium. Carnotite commonly occurs disseminated throughout weathered sedimentary rocks, especially sandstone. California has been one of the largest producers of carnotite. It has also been found in Australia and elsewhere.

(8) Non-metallic or Industrial Minerals

Included here are those minerals that either do not contain metals or are not used for their metal content. It is in this category that such valuable and varied materials as coal, petroleum, sulphur, fertilizer, building stones, and gem stones are placed. Some of these products and the minerals from which they come are described below.

Abrasives. Materials used to polish, abrade or cut other materials are called abrasives. Minerals commonly used for this purpose are garnet, diamond, corundum and varieties of quartz.

Asbestos. Certain of the fibrous silicate minerals are useful in such things as insulating, fireproofing, the manufacture of plastic and in brake linings. The most important of these are chrysotile, crocidolite and actinolite–tremolite.

Cements, Lime, and Plasters. Limestone (see page 59), composed primarily of calcium carbonate, is used in the manufacture of Portland cement, agricultural lime and as a building stone. It is also very important in the steel industry. **Gypsum,** calcium sulphate, is used in making plaster board, paints and plaster of Paris.

Clays. Clay minerals, in combination with certain other minerals, provide the basic raw materials for the brick, tile, pottery and china industries. Clay minerals are also used in making paper, linoleum, cement, and in foundry work. Certain types of clays are also used as refractory brick to line kilns and furnaces.

Mineral Fertilizers. The three essential elements for promoting plant growth are potassium, nitrogen and phosphorus. Phosphate rock is a valuable source of phosphorus because it contains a large amount of the mineral apatite. Sylvite is an important source of potassium, and natural nitrates are used in mineral fertilizers to provide nitrogen. Most of the latter are mined or quarried in Chile. Ground limestone, glauconite, gypsum, borax and, increasingly, anhydrite are also used in certain mineral fertilizers.

Salt. Halite, common rock-salt, is much used in the chemical industry as an important source of sodium compounds and (to a lesser degree) chlorine. It is also employed in the tanning of leather, food preparation and certain refrigerants. These are but a few of the many uses of this all-important mineral which has long been so valuable to man.

Sulphur. This yellow, non-metallic mineral is found in volcanic rocks, around hot springs and associated with salt domes. Sulphur is produced principally from the 'cap rocks' of salt domes in Texas and Louisiana in the United States. Lesser amounts come from Sicily and elsewhere.

Sulphur and its compounds are used in making paper, sulphuric acid, gunpowder, matches, insecticides and medicines. It is also used in the process of vulcanizing rubber. The spa waters of Harrogate, Cheltenham and Bath contain sulphuretted hydrogen (H_2S), a gas which smells like bad eggs. Along with the other mineral contents of the 'waters', it is thought to have a beneficial effect on the tired and elderly.

CHAPTER THREE

IGNEOUS ROCKS AND VOLCANISM

Igneous rocks are those rocks that have solidified from an original molten state. The word igneous is derived from the Latin word *ignis*, meaning 'fire'. Temperatures deep within the earth are exceedingly high and many rocks and minerals exist in a molten condition called **magma**. Magmas are large bodies of molten rock deeply buried within the earth. Sometimes magmatic materials are poured out upon the surface of the earth, as, for example, when lava flows from a volcano. These molten materials which spill out upon the surface are known as **eruptive, extrusive** or **volcanic** rocks. Under certain other conditions, magmas do not come to the surface but may force their way or intrude into other rocks where they solidify. These intruding rock materials harden and form **intrusive or plutonic** rocks.

Igneous rocks may be distinguished from sedimentary rocks (Chapter Four) and from rocks (Chapter Five) by their texture and structure, their mineral content, and their complete lack of fossils.

(1) Intrusive or Plutonic Rocks

These are rocks that have solidified from molten mineral mixtures beneath the surface of the earth. The more deeply buried intrusive rocks tend to cool slowly and develop a coarse texture composed of

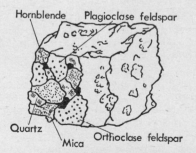

Fig. 20. Granite, a coarse-textured intrusive igneous rock

relatively large mineral crystals (Fig. 20). On the other hand, those that cooled more quickly (because they were nearer the surface) are finer textured. The texture of an igneous rock depends largely on the shape, size and arrangement of the grains comprising it. As a

39

result of the crowded conditions under which mineral particles are formed, they are usually angular and irregular in outline. Some typical intrusive rocks (granite, gabbro, peridotite, syenite and diorite) are described below.

Granite. Granite (Fig. 20) and its associated forms, granodiorite, aplite, pegmatite and microgranites, are some of the commonest and best known of the coarse-grained intrusive rocks. Generally the crystals are all between about $\frac{1}{16}$ inch and $\frac{1}{2}$ inch across, but some granites show what is called 'porphyritic' structure in which a number of large crystals (phenocrysts) stand out clearly. For example, Shap granite shows conspicuous pink phenocrysts of orthoclase and is therefore easily recognizable. Large boulders of Shap granite are to be found on the beach of Robin Hood's Bay, Yorkshire, some 80 miles from Shap, where they had been brought as erratics (see page 108) by the ice. These crystals, however, are dwarfed by the gigantic ones (20 or more feet across) which have been found in pegmatites in Norway and in micas in South Africa.

Essentially granites are composed of quartz, feldspars and mica or hornblende. **Aplites** contain a large percentage of orthoclase feldspar and are found in granitic veins, as are **pegmatites** which are a much coarser grained form, also found in dykes and in the country rock. **Microgranites** are fine-grained granites sometimes found in the margins of plutonic intrusions; they may also be found in extrusive lavas. When the biotite (dark-coloured mica) content of a granite is increased the rock is called **granodiorite**.

Granite, which is typically light in colour, may be white, grey, pink or yellowish brown. Although its individual grains are clearly visible they are well cemented together and form a hard, enduring rock. It can also be polished and is therefore used as a building stone and in monumental masonry. Two of the commonest forms used are Shap granite and that of Dartmoor.

The origin of granites is a subject on which geologists hold varying views. Many were formed by cooling and solidifying of the magma at great depth: they are therefore plutonic igneous rocks, formed by selective melting of the Sial (see page 144). It is significant that granites are restricted to continental regions. Granites are often associated with metamorphic rocks and it is also known that any rock may be altered to a granite given sufficient time, a process called **granitization**. It is thought that, by the action of heat and pressure, the rocks of the crust became virtually molten and the metamorphic granites were created. A great deal of controversy has arisen over whether granites are igneous or metamorphic, and it is now generally accepted that there are two forms.

Gabbro. Gabbro is a heavy, dark-coloured igneous rock consisting of coarse grains of plagioclase feldspar and augite. Quartz is absent, (except in quartz-gabbro), and the minerals crystals are usually dark grey, dark green or black.

Peridotite. This rock, in which the dark minerals are predominant, is also called a pyroxenite. Kimberlite, a peridotite composed of a pyroxene-olivine mixture, is famous for the large numbers of diamonds that have been extracted from it. As its name suggests, kimberlite is found in Kimberley, South Africa.

Syenite. Syenite resembles granite, but is less common in its occurrence and contains little or no quartz. If quartz is present the rock is referred to as a quartz-syenite. Consisting primarily of potash feldspars with some mica or hornblende, syenites are typically even-textured and the mineral crystals are usually small.

Diorite. Diorite is often found in dykes stemming from granite masses and is coarse-grained. Apart from the absence of quartz, diorite is very much like granite.

(2) Extrusive or Volcanic Rocks

Extrusive igneous rocks are formed when molten rock solidifies after forcing its way out to the surface of the earth. Such rocks may pour out of the craters of volcanoes, or from great fissures or cracks in the earth's crust. In addition to the liquid lava, solid particles such as volcanic ash or volcanic 'bombs' may also be thrown out during eruptions.

As the magma reaches the surface, it loses its gases and undergoes relatively rapid cooling. This prevents slow crystal growth and re-sults in a microcrystalline texture in which the crystals cannot be seen with the unaided eye. Sometimes the magma cools so rapidly that no crystallization occurs, and this produces volcanic glass.

Some of the more common extrusive rocks are felsite, basalt, pumice and obsidian.

Felsite. Felsite is a general term applied to igneous rocks of very fine texture, including the glassy **rhyolites** which grade into pitchstone and obsidian, and **trachytes** and **andesites** similar in composition to syenites and diorites, respectively. So fine is the texture that the individual grains, which are light in colour, cannot be seen through an ordinary magnifying glass. Felsites may range in colour from white to light- and medium-grey, and to shades of pink, red, green, purple or yellow. Felsite commonly contains quartz, orthoclase feldspar and biotite mica. These are the same minerals found in granite, a typical intrusive rock. Granites, however, are coarsely crystalline in texture, and the felsites are typically fine-grained. We see, then, that the same magma (containing quartz, orthoclase feldspar and biotite mica), cooling at different distances from the surface, may form rocks of similar chemical composition, although their physical appearance is quite different.

Basalt. This is one of the world's most abundant extrusive rocks. Basalts are typically dark grey, dark green, brown, or black in colour, and are normally quite heavy. They are fine grained in texture and consist primarily of pyroxene, plagioclase feldspar, and, in some cases, olivine. Some basalts are characterized by a large number of open spaces or pores which mark the site of former gas bubbles. This porous rock, called **scoria**, is common in many hardened lava flows. With the passage of time the pores, or vesicles, may become filled with minerals such as quartz or calcite. Such mineral-filled vesicles, usually almond-shaped, are known as **amygdales.** Basalts characterized by large numbers of amygdales are called amygdaloidal basalts, and often yield fine mineral crystals.

Basaltic rocks may be seen in various parts of Britain, including the Midland Valley of Scotland where they are of Carboniferous age.

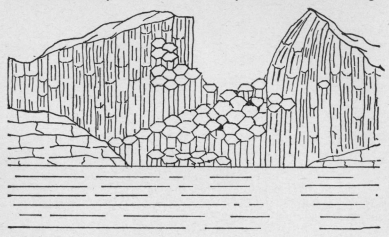

Fig. 21. Columnal jointing in basalt (Giant's Causeway, Antrim)

In India and the north-western United States there are large basalt flows covering about 200,000 square miles (512,000 km²) to depths of thousands of feet.

Basalt commonly displays columnar jointing. This phenomenon, which comes about as the rock cools, shrinks and splits into vertical columns, can be seen most dramatically in the Giant's Causeway in County Antrim, Northern Ireland, and Fingal's Cave on the Island of Staffa in the Hebrides.

Because of their hardness, basalts are valuable for road building and other construction purposes. In addition, large deposits of copper have been found in amygdaloidal deposits.

Fig. 22. Pumice, a type of intrusive igneous rock

Pumice. Lava that solidified while steam and other gases were still bubbling out of it is called pumice. It is formed from a rapidly cooling volcanic froth and is characterized by the presence of large numbers of fine holes which give the rock a spongelike appearance (Fig. 22). Pumice is very light in weight and, because many of the air spaces are sealed, it can float on water. Blocks of pumice thrown into the sea during eruptions of island volcanoes have been known to float for great distances from their site of origin. Pumice is typically light in colour and, though differing greatly in appearance, has the same chemical composition as obsidian and granite. Pumice is found wherever there has been volcanic activity throught the world.

It is used as an abrasive, in soap, cleansers, and some rubber erasers.

Obsidian. Known also as volcanic glass, obsidian is a glassy extrusive rock which cooled so rapidly that there was no formation of separate mineral crystals. It is a lustrous, glassy, black or reddish-brown igneous rock (Fig. 23). Obsidian exhibits a conchoidal fracture

Fig. 23. Obsidian

which leaves sharp edges; consequently this stone was commonly used by early man to make arrowheads, spear points, knives, and other implements. Formed by the rapid cooling of surface lava flows, obsidian occurs in various parts of the world. The two best known sites are Mount Hecla in Iceland, and Yellowstone National Park in Wyoming (where it was sought by the Indians for use in spearpoints and arrow-heads).

(3) Texture of Igneous Rocks

Texture is a physical characteristic of igneous rock that is influenced by the rate of cooling or crystallization of a magma. Commonly used to refer to the general appearance of any rock, it refers more specifically to the shape, size and arrangement of silicate minerals in the igneous rocks. A rock with mineral grains large enough to be seen and identified with the unaided eye is said to have a **granular**, or **granitic**, texture. If the individual mineral grains in a rock are too small to be seen with the naked eye, the texture of the rock is said to be **aphanitic.** Glassy-textured rocks such as obsidian appear to be composed of glass. Some igneous rocks appear to have a mixed texture. This type of texture is called **porphyritic** and is characterized by relatively large crystals, called **phenocrysts,** surrounded by a **groundmass** (or background) of smaller crystals

(Fig. 24). **Porphyries,** as such rocks are called, are believed to represent two distinct phases of cooling and solidification.

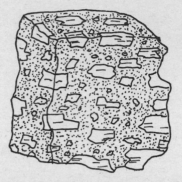

Fig. 24. Porphyry showing light-coloured phenocrysts in the darker groundmass

Intrusive rocks which cool at a very slow rate may form crystals ranging from several inches to several feet in length. Such coarse grained rocks are called **pegmatites** and are especially characteristic of certain types of granite.

(4) Chemical Composition of Igneous Rocks

The type of igneous rock that is formed by a magma or lava is primarily dependent upon the chemical composition of the original molten rock material.

High-silica-content or **Acidic** Igneous Rocks. Those igneous rocks that have a high silica content are known as acidic or **sialic** rocks. The term 'sialic' is derived from the chemical symbols Si (for silicon) and Al (for aluminium). Such rocks have a high content of silica and sodium-potassium feldspar. They typically contain relatively low percentages of iron, magnesium, and calcium. Acidic igneous rocks are usually light in colour and have a low specific gravity. They are the predominant rocks of the continents and are represented by such common examples as granite, rhyolite, and pumice.

Low-silica-content or **Basic** Igneous Rocks. These are dark-coloured, relatively heavy igneous rocks. They are known as basic or **simatic** (from *si*licon and *ma*gnesium) rocks because of their low silica content and proportionately higher content of such iron-magnesium minerals as biotite, olivine, pyroxene, and hornblende. Basic rocks underlie the acidic rocks of the earth's crust, and are believed to make up most of the volcanic islands and to form large parts of the deep ocean floor. Gabbro and basalt are typical basic igneous rocks.

There are numerous gradations between acidic and basic rocks, and many specimens fall into an intermediate or transitional category.

(5) Igneous Rock Formations

Intrusive (plutonic) igneous rocks have been intruded or injected into the surrounding rocks. Intrusions of this type normally occur at great depth, consequently igneous intrusive bodies may be seen only after the overlying rocks have been removed by erosion. Some of the more common intrusive bodies are discussed below.

Dykes. A dyke is a tabular or wall-like mass of igneous rock that cuts across bedding planes (see page 57) when introduced into sedimentary rocks (Fig. 25). Dykes commonly result from magma being

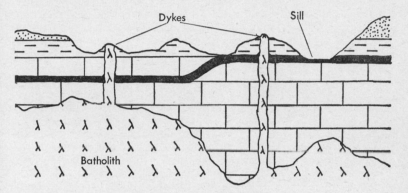

Fig. 25. Igneous intrusions

injected into cracks and joints in the rocks, and range in size from a few feet to many miles in length. They are quite common in volcanic areas, and are usually associated with volcanic necks.

Probably the best known dyke in Britain is the Cleveland dyke in Yorkshire. It is composed of dolerite and is over 30 feet thick. In some places, e.g. the Isle of Arran off the west coast of Scotland, the dykes occur in swarms. Elsewhere, e.g. in Mull (also off the west coast of Scotland), they form ring-dykes around a large igneous mass.

Sills. These are tabular bodies of igneous rocks that spread out as horizontal sheets between beds or layers of rocks. They differ from dykes in that the igneous rocks lie parallel to the bedding plane. The best known sill is the Whin Sill, running across northern England and southern Scotland, which Hadrian used as a foundation for part of his Wall. Others are found in the Isle of Arran associated with the dykes. Probably the largest sill is the 1,000-ft. sill which gives rise to the Palisade Cliffs overlooking New York.

Laccoliths. Laccoliths are lens-like or mushroom-shaped intrusive bodies that have relatively flat under-surfaces and arched or domed upper surfaces. They are intruded between the bedding planes, and differ from sills in that they are thicker in the centre and become thinner near their margins. Such an intrusion may give rise to a dome-shaped hill: Corndon in Shropshire, for example.

Batholiths. These, the largest of igneous intrusions, are irregularly shaped and may cover thousands of square miles. Batholiths extend great distances within the earth and become larger with depth. Dartmoor, the Cairngorms, Scotland and Mourne mountains, Ireland, are examples of places where batholiths have been exposed at the surface.

Stocks. These are similar to batholiths, but cover an area of less than 40 square miles. It is thought that Bodmin Moor, Land's End and the Scilly Isles are stocks off the same huge batholith that gives rise to Dartmoor.

Volcanic Neck. A volcanic neck is formed when the lava-filled conduit of an extinct volcano is exposed by erosion. Necks are usually less than a mile in diameter, and as they are often more resistant than the rocks which originally surrounded them, they may stand up as spires or columns. Volcanic necks abound in Scotland: for example, Castle Rock in Edinburgh, North Berwick Law in East Lothian, Inchcape Rock and Bass Rock.

Extrusive Formations. The most common features formed by extrusive igneous rocks are lava flows, generally of a sheetlike nature. Certain lava flows, such as those of the Columbian Lava Plateau between the northern Rocky Mountains and the Cascade Range in the north-western United States, cover hundreds of square miles to a depth of almost a mile. Lava flows are also common in other parts of the world, including India (Bombay), Hawaii and Britain.

It may appear strange that volcanic necks, lava flows and other evidence of volcanic activity should be found in Britain which does not lie in a volcanic zone (see Fig. 27), but these zones only represent the situation as it is today. In the past Britain has often experienced volcanic activity. The volcanic features mentioned above were formed over 300 million years ago in what we know as the Carboniferous Period. The last time Britain experienced volcanic activity was about 50 million years ago during the Tertiary Period, when the extensive basalt flows of Northern Ireland and the Scottish Isles were extruded. (We shall learn more about the continually changing earth as we look at Historical Geology in Chapter Eighteen.)

Some lava flows are associated with volcanoes; others are the result of fissure flows. Many of these flows exhibit a typical columnar jointing (Fig. 21). Others consist of rough, massive blocks of scoria.

Extrusive igneous products other than lava flows include volcanic ash, dust and 'bombs'. Volcanic bombs are spherical or pear-shaped objects (Fig. 28) formed when large masses of lava harden as they swirl through the air.

(6) Volcanoes

Volcanoes are vents in the earth's crust through which molten rock and other volcanic products are extruded. Volcanoes and their related phenomena, such as **fumaroles** and **hot springs,** are among the most interesting of all geologic processes, and their activities have been mentioned in the earliest historical writings.

The typical volcano is a cone-shaped mountain with a funnel-shaped crater at the top (Fig. 26). This crater is connected with the underground magma chamber by means of a central vent or conduit, and during periods of eruption, steam, dust, ashes, stones, and molten

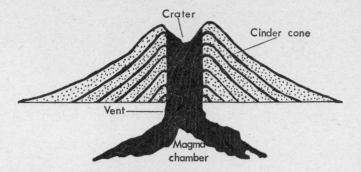

Fig. 26. Volcanic cone with crater

rock (lava) emanate from the vent. The magma chamber is located far beneath the surface of the earth and is a reservoir containing hot molten rock material, which may be either intruded into the earth's crust or extruded upon the surface.

Distribution of Volcanoes. The volcanoes of the earth appear to be concentrated within certain well-defined geographic belts or zones (Fig. 27). These volcanic zones occur most frequently in areas

Fig. 27. Volcanic zones around the world

of crustal instability within or near regions of recent mountain-building activities. The two major zones are associated with large fault, or fracture, zones in the earth's crust.

The Pacific Zone is the more important of the two major zones, and is located along the border of the Pacific Ocean. This zone includes the volcanoes of South and Central America, Alaska, the Aleutian Islands, Japan, the Philippines, and the East Indies.

The Mediterranean Zone extends in an east–west direction and includes the volcanoes of the Mediterranean basin, the West Indies, Hawaii and the Azores.

In addition to these two major zones, there are also volcanoes in the Atlantic, Pacific, and Indian Oceans, in Iceland, and in the Antarctic.

Activity of Volcanoes. Some volcanoes are active, while others have not erupted within historic times. To indicate the relative activity of volcanoes, they have been divided into three classes: active, dormant, and extinct. Those which are continually or periodically in a state of eruption are classified as **active** volcanoes. A volcano that is now inactive, but has been known to erupt within the last thousand or so years is classed as **dormant.** There are many examples of 'dormant' volcanoes, such as Vesuvius, experiencing violent eruptions after many centuries of inactivity.

Vesuvius, near Naples in Southern Italy, is famous for its violent eruption in A.D. 79 which buried the Roman cities of Pompeii and Herculaneum. Another great eruption in 1906 reduced the height of the mountain by several hundred feet.

An **extinct** volcano is one which is not known to have erupted within historic times, i.e. the last four or five thousand years. But here again Nature has a tendency to disregard man-made classification, for 'extinct' volcanoes have been known to up-grade themselves without warning to the 'active' class. For example, Lassen Peak in northern California erupted in 1914 after a 200-year period of inactivity. During this eruption clouds of steam and ash issued from the crater, and one column of steam rose to a height of 10,000 feet above the crest of the mountain. Its unexpected activity is believed to have been the result of great earthquakes in Alaska and California which preceded the 1914 eruption. The peak, known also as Mount Lassen, is surrounded by lava flows; excellent examples of volcanic rocks, hot springs, mud pots and other activities related to volcanism may be seen in this area. (A mud pot is a type of hot spring consisting of a shallow pit filled with hot, normally boiling, mud and very little water.) Volcanologists tell us that only those volcanoes that have been eroded almost to the level of their magma chambers can be classed as truly extinct.

(7) Volcanic Products

When volcanoes erupt they may eject a large variety of material ranging from gases to large fragments of rock.

Gases. The gases that are emitted from volcanoes are composed largely of water vapour, with varying amounts of carbon dioxide, hydrogen sulphide and chlorine. During an eruption, these escaping gases may become mixed with vast quantities of volcanic dust, and often rise from the crater in great dark clouds which may be seen for many miles.

Liquids. The liquids produced by volcanoes are the lavas—great quantities of white-hot molten rock. Lava is more typically extruded from the crater in the top of the volcano, but it is not uncommon for the lava to break through the sides of the cone and escape by means of fissures developed along zones of weakness.

Not all lavas are alike in their chemical and physical properties, and these properties may be reflected by the manner in which the volcano erupts. The chemical composition of the lava will affect its viscosity, which will in turn affect its rate and distance of flow. Chemical composition will also affect the shape of the cone, and will have some bearing on the surface structure of the rock formed when the molten rock solidifies.

Because of the different characteristics of lavas, geologists have classified them as acidic, basic, and intermediate. The **acidic lavas** are high in silica content (65 to 75 per cent), usually very viscous, and frequently explosive. **Basic lavas** are low in silica (less than 50 per cent), less viscous, and not as likely to be explosive, because the dissolved gases can escape more easily from the more fluid lava. **Intermediate lavas** are those that fall in between the acidic and basic categories: a typical intermediate lava contains 50 to 60 per cent silica.

The composition of the lava and the manner in which it cools and solidifies are often reflected in the surface structure of the rock. As the lava flows on to the earth's surface, it cools and there is a reduction of pressure which permits the trapped gases to escape. These escaping gases produce bubbles which leave empty vesicles when the lava cools. Hardened cindery lava which contains large numbers of irregular holes is called scoria. A lava surface covered with jagged angular blocks of scoria s called *aa* (pronounced AH-ah), and lavas with relatively smooth billowy or ropey surfaces are called *pahoehoe* (pah-HOE-ay-HOE-ay). These strange-sounding terms originated in the Hawaiian Islands, where both of these forms of lava occur.

In various parts of Britain there are **pillow lavas,** which were formed when sheets of lava were extruded from underwater volcanoes. As the rolling hot lava met the cold water its surface was hardened so that further extension of the flow was prevented. The lava then seeped through cracks and solidified in pillow-like heaps Those found in North Wales date from the Ordovician period, but similar examples are being formed by submarine volcanoes today.

Solids. The solid matter ejected from volcanoes may vary in size from fine dust to huge blocks of rock weighing several tons. These solid products are referred to as **pyroclastic** material, and include the volcanic 'bombs' (rounded objects formed by the rapid cooling of molten lava as it is hurled through the air) shown in Fig. 28. The large, angular fragments of rock ejected from a volcano are known as volcanic **blocks;** the smaller pieces (about the size of a pebble) are called **lapilli.** Great quantities of volcanic cinders and volcanic dust may be produced when small particles of lava solidify after having been thrown into the air.

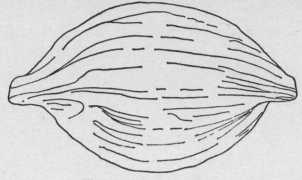

Fig. 28. A volcanic bomb

(8) Volcanic Eruptions

Volcanic eruptions may be classified as either central eruptions or fissure eruptions. In a **central eruption** the volcanic material is ejected through a single conduit or vent that opens into a crater at the top of a volcanic mountain. Volcanic products erupted in this manner commonly build a cone which may be composed of lava, cinders, or alternating layers of both. An eruption of this type may be either explosive or quiet, depending on the physical and chemical characteristics of the lava. If acidic lavas are being ejected they are apt to be more explosive, while the basic lavas will normally flow out relatively quietly.

Fissure eruptions occur when large quantities of lava are extruded through a large fissure, or series of fissures, in the earth's crust. Eruptions of this type may affect wide areas and are believed to be responsible for the great lava plains and basalt plateaus of the world. Most of these lava plains were formed in prehistoric times, and volcanic action is no longer present in the immediate vicinity of the flows. The only fissure eruption within modern times occurred in Iceland in 1783 when a flood of lava began pouring out along a large fissure near Mount Skapta. Known as the Laki fissure, it eventually covered an area of 218 square miles (570 km²). However there is much evidence to suggest that large fissure eruptions have occurred at many times during the geologic past. The enormous volumes of lava which issue from long fissures produce sheets much more widespread than those from a central cone. The Tertiary flows in Britain and Iceland, and the great Indian basalt flow, may well be the results of fissure eruptions.

(9) Types of Volcanoes

One of the more commonly used classifications of volcanoes recognizes four principal types: Pelean, Vulcanian, Strombolian and Hawaiian.

Pelean. The Pelean or explosive type of volcano erupts with violent explosions which eject great quantities of gas, volcanic ash, dust and large rock fragments. Such eruptions are believed to occur in volcanoes whose vents have been plugged by solidified magma. The gases accumulating in the magma chamber eventually develop pressures great enough to blow out this plug, and the resulting explosion is often so violent that a large part of the mountain is completely blasted away. Eruptions of this type are often accompanied by destructive clouds of volcanic gases and dust, and are among the most disastrous kinds of explosion known to man. An example occurred in 1902 at Mount Pelée on the Island of Martinique in the West Indies. After eruptions in 1762 and 1851, the volcano gave no sign of life and was thought to be dormant; but in 1902 it erupted with such violence that the top of the mountain was blown off. This eruption was accompanied by a huge black cloud of hot gas and dust which descended on the town of St. Pierre and caused the deaths of about 30,000 people.

Vulcanian. The Vulcanian type of volcano is characterized by very viscous lavas, which solidify soon after they come in contact with the air. The surface of the lava in the crater solidifies between eruptions, and each later eruption takes place through breaks in this crust. This kind of eruption typically produces large quantities of ash, lava and great clouds of dust and gas. Mount Vesuvius in Italy is noted for its great numbers of eruptions followed by periods of quiet, and is a typical example of the Vulcanian volcano.

Strombolian. This type of volcano is in constant action, with the pent-up gases escaping spasmodically—unlike the more typical volcano which has alternating periods of activity and quiescence. Stromboli, an island of the Lipari group in the Mediterranean Sea, off the northern coast of Sicily, is the classic example and gave the type its name. It is perpetually active, and explosive eruptions come at intervals of a few minutes to an hour or so. These explosions are accompanied by the emission of viscous lavas and large amounts of pyroclastics. Both the Strombolian and Vulcanian types of volcanoes have also been referred to as **intermediate** volcanoes.

Hawaiian. The Hawaiian, or quiet, volcano is characterized by less viscous lavas which permit the escape of gas with a minimum of explosive violence. The lava is typically extruded from the crater, although flank eruptions may occur through fissures in the side of the mountain. Eruptions are commonly accompanied by small explosions of escaping gas, which may whip the lava into a froth that later solidifies into scoria. Probably the best known volcano of this type is Mauna Loa in the Hawaiian Islands. This volcanic mountain rises 13,680 feet above sea level, and its oval-shaped crater is about five miles in circumference. The walls of this great crater are almost vertical and it is believed to be about 1,000 feet deep.

(10) Land Forms Produced by Volcanic Activity

Volcanic activity and the extrusion of lava result in four principal types of land forms: plateau basalts (lava plains), volcanic mountains, volcanic craters, and calderas.

c

Plateau Basalts or **Lava Plains.** These are formed when floods of lava are released by fissure eruptions and spread in sheetlike layers over the earth's surface. The three largest basalt plateaus are the Columbia River Plateau, 200,000 square miles (520,000 km²) in extent and nearly a mile thick, the Deccan Plateau of India, and the Parana Plateau of South America.

Volcanic Mountains. These are mountains that are composed of the volcanic products of central eruptions, and are classified as **explosion cones** (or cinder cones), **composite cones** (or strato-volcanoes), and **lava domes** (or shield volcanoes).

Explosion cones are formed solely by explosive eruptions and consist of successive, steeply inclined layers of pyroclastics disposed around a central crater. Cones of this type seldom exceed 1,000 feet in height and are often the result of a single volcanic explosion.

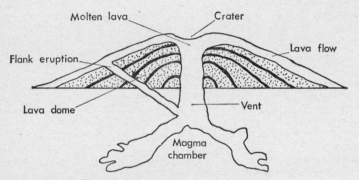

Fig. 29. Lava dome, or shield volcano

Composite cones are steeply sloping volcanoes composed of alternating sheets of lava and pyroclastic material. They are cone-shaped mountains with concave sides and may be as much as 12,000 feet high. The alternating layers of lava and pyroclastics are evidence of intermittent periods of quiescence and explosive eruption. Some famous composite cones or strato-volcanoes are Vesuvius in Italy, Fujiyama in Japan, Mount Rainier at Washington, and Kilimanjaro in Tanzania.

Lava domes are broad, gently sloping volcanic domes, characteristically exhibiting a gently rounded convex upper surface (Fig. 29)—hence the alternative name of shield volcanoes. This type of volcanic mountain is composed of large numbers of overlapping basaltic lava flows which originated from the central vent or from flank eruptions through fissures. The great volcanoes of Hawaii are typical of this type of mountain.

Volcanic Craters. These are funnel-shaped depressions in the tops of volcanic mountains through which central eruptions take place. Most craters have been produced as a result of explosive volcanic activity and are seldom more than a mile in diameter or more than a few hundred feet deep.

Calderas. Calderas are nearly circular, basic-shaped depressions in the tops of volcanoes, and are much larger than craters. There are two types of calderas: one is brought about as the result of explosive activity; the other is a result of collapse or subsidence. Explosion calderas are created by violent volcanic explosions which have displaced large amounts of rock. Collapse or subsidence calderas are produced when the upper part of the volcano collapses because of the sudden withdrawal of the supporting magma. Some calderas are believed to have been formed by the combined effects of explosion and collapse. One example is Crater Lake in Oregon, U.S.A., which occupies a great caldera and may have been formed by the collapse of a now-extinct volcano. The caldera is filled with water which covers an area of approximately 20 square miles and is as much as 2,000 feet deep. Wizard Island is a small cinder cone rising about 800 feet above the lake. This cone was formed by eruptions that occurred after the top of the volcano, Mount Mazama, was destroyed.

(11) Sources of Volcanic Heat

Although the ultimate cause of volcanism is not definitely known, several attempts have been made to explain it. Among the most acceptable of these ideas are the pressure-release theory, the frictional-heat or compression theory, and the radioactivity theory.

The Pressure-Release Theory. This theory assumes that the greater the depth into the earth, the higher the temperature; therefore the very deeply buried rocks are potentially liquid. However, these rocks are under such great pressure that they must remain in the solid state except in those areas where breaks in the earth's crust allow them to liquefy as a result of diminishing pressures.

The Frictional-Heat or **Compression Theory.** In this theory it is assumed that the heat is generated by friction which occurs during deformation of the earth's crust (see Chapter Five). This has been suggested because of the proximity of areas of volcanic activity to regions of relatively recent crustal deformation.

The Radioactivity Theory. This theory postulates that local concentrations of radioactive materials are capable of generating enough heat to melt the large amounts of rock necessary for volcanic activity.

(12) Fumaroles, Hot Springs and Geysers

In many areas of volcanic or other igneous activity there is evidence of volcanic gases, steam or hot water escaping from the earth. Some of these phenomena are discussed below.

Fumaroles. These are vents or cracks in the earth's surface through which steam and gas issue. Steam emanates from certain fumaroles in Italy in sufficient quantities to drive turbo-electric generators. Some fumaroles are characterized by the emission of large quantities of sulphurous vapours and these have been called solfataras.

Hot Springs. Hot springs are formed when ground water is heated by large masses of magma located relatively near the surface, such as those at Bath and other spas. These occur in other parts of the world, as at Whakarewarewa in New Zealand, where they have been used for centuries by the Maoris as natural 'cookers'.

Geysers. A geyser is a special type of hot spring that intermittently erupts a column of steam and hot water. Geysers originate in areas where ground temperatures are unusually high, and where long narrow fissures are likely to be present in the rocks. The ground water at the bottom of these fissures is heated to a temperature far above the boiling point of water (100°c). This superheating of the bottom water is made possible because of the pressure exerted by the water that lics above it. As the water in the bottom of the fissure expands, it causes some of the overlying water to overflow on to the surface. This overflow relieves part of the pressure and allows the superheated water to explode into steam, ejecting a great column of water into the air. Some geysers, such as Old Faithful in Yellowstone National Park, erupt with amazing regularity, but most geysers are quite erratic in their performance. The Yellowstone region of Wyoming in the United States is famous for its hundred or so geysers and 3,000 non-eruptive hot springs. Other areas where geysers occur are Iceland, New Zealand and Japan.

(13) Recent Volcanic Activity

After 200 years as a dormant volcano, Krakatoa, located in Sunda Strait between Java and Sumatra, erupted in 1883 to produce probably the biggest explosion in history. This great blast threw debris many miles into the air, and falling ashes were distributed over 300,000 square miles (800,000 km^2); in 15 days volcanic dust from the explosion had encircled the earth. The violent activity of Krakatoa was responsible for the generation of huge tidal waves, some as much as 100 feet in height, which destroyed hundreds of villages and drowned an estimated 36,000 people.

Paricutin, one of the most recent and most publicized of volcanoes, has provided us with much information about the birth, development and death of a volcano. Located about 200 miles west of Mexico City, it first showed signs of activity in February 1943 when extensive lava flows and explosions of pyroclastic material were reported. A 350-foot cinder cone had been formed by the end of the first week's activity, and this cone attained a height of about 1,400 feet in the first year. Eruptions ceased in February 1952, and Paricutin now appears to be dead.

The most violent eruption of this century occurred in March 1956 on the peninsula of Kamchatka in the extreme east of the Soviet Union. Volcanic activity had restarted six months earlier, and a team of Russian scientists who had gone to the area were able to observe and document the remarkable events that followed. It appears that the longer a volcano remains dormant the greater is the explosion when it does erupt. This region had been dormant for at least 5,000 years, so the explosion of the pent-up material was of gigantic proportions. Gas and debris (enough to cover a city the size of Birmingham with a 50-foot layer) shot up over 100,000 feet into the air.

Part of the team, camping only a few miles from the crater, were almost deafened by noise and felt the ground vibrating beneath their feet. At this time the crater was a mere 800 feet across; by the time the activity ceased a crater of over 5,000 feet in diameter had been formed. Another part of the team was 25 miles from the volcano and was caught in a shower of ashes which lay on the snow to a depth of several inches. The dust cloud spread, and fine ash fell around the North Pole. Four days after the explosion a small dust cloud was noted over Britain.

The above examples are just three of the many sensational eruptions that have been recorded. The list of suggested further reading at the end of this book gives details of where other accounts of famous volcanic activity can be found.

In discussing pillow lavas on page 49, mention was made of submarine volcanoes. Often erupting at great depth, these do not make such a spectacular show as terrestrial volcanoes. But in 1957 an example occurred in the Azores, where the sea was only 300 feet deep. An observer watching through binoculars noted a curious bubbling in the sea. Soon the sea turned a dirty yellow colour as gases escaped from the erupting volcano. Within a few hours the sea in the area boiled and a column of vapour shot into the air. The pumice extruded by the volcano was very light and floated as a brown mass on the surface. Within a week or two the accumulating material had built an island over 300 feet high and 2½ miles in diameter. It did not remain an island for long, for it was soon connected to neighbouring land by a peninsula of debris a mile wide. Unlike the tremendous din of the Kamchatka eruption all this took place in near silence, the noise being muted by the sea. Similar cases of islands suddenly erupting occasionally make the newspaper headlines.

CHAPTER FOUR

SEDIMENTARY ROCKS

Those rocks which are exposed on the earth's surface are especially vulnerable to the agents of erosion. They may be attacked chemically, or they may be mechanically broken and worn by, for example, being rolled along the bottom of a stream. These rock fragments are commonly picked up and transported by wind, water, or ice, and when the transporting agent has dropped them they are generally referred to as **sediments.** Sediments are typically deposited in layers or beds called **strata.** When sediments become compacted and cemented together (a process known as **lithification**), they form sedimentary rocks.

These rocks, represented by such common types as sandstone, shale, and limestone, make up about 75 per cent of the rocks exposed on the earth's surface. The majority of them have been formed in marine conditions, or at least are associated with water in one way or another.

Sedimentary rocks are generally classified as either **clastic** or **chemical,** according to the source of the rock materials which form them.

(1) Clastic Sedimentary Rocks

Clastic sediments are composed of rock fragments which have been derived from the decomposition or disintegration of igneous, sedimentary, or metamorphic rocks. Rocks formed from these worn-down rock particles are also called **detrital** or **fragmental** sedimentary rocks. Because the sediments which form these rocks are normally transported by mechanical means (wind, water, or ice), they have also been referred to as **mechanical** sediments.

TABLE 2 Classification of clastic rocks (rounded, sub-rounded and sub-angular)

SIZE	FRAGMENT	AGGREGATE
Over 256 mm	Boulder	Boulder gravel, boulder conglomerate
64–256 mm	Cobble	Cobble gravel, cobble conglomerate
4–64 mm	Pebble	Pebble gravel, pebble conglomerate
2–4 mm	Granule	Granule sand
$\frac{1}{16}$–2 mm	Sand	Sand, sandstone (*arenaceous rocks*)
$\frac{1}{256}$–$\frac{1}{16}$ mm	Silt	Silt, siltstone (*argillaceous rocks*)
Under $\frac{1}{256}$ mm	Clay	Clay, shale

Clastic rocks are composed of rock particles of varying size. We use the terms **arenaceous** (sandy) for clastic rocks with grain size between $\frac{1}{16}$ and 2 mm and **argillaceous** (clayey) for those less than $\frac{1}{16}$ mm. Table 2 shows the most generally accepted size ranges of the materials composing clastic sedimentary rocks. Some of the more common types of clastic sedimentary rocks are described below.

Shale. The most abundant of all sedimentary rocks, shale is formed from silt and clays which have hardened into rock. Shale (Fig. 30) is characteristically fine-grained, thinly bedded, and split easily along bedding planes (the dividing planes which separate individual layers of beds of sedimentary rocks). Shale containing appreciable amounts of sand is called **arenaceous** shale; that containing large

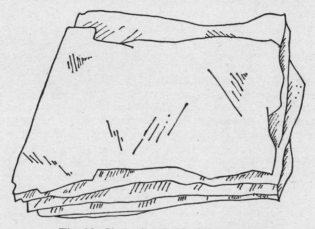

Fig. 30. Shale showing bedding planes

amounts of clay is said to be **argillaceous. Carbonaceous** shale is typically black, and high in organic matter; shale which contains large amounts of lime is known as **calcareous** shale. Carbonaceous shale may yield petroleum or coal, and calcareous shale (or clay) is used in the manufacture of Portland cement. **Oil** shales, found in central Scotland, are generally blackish in colour because of the presence of organic matter—including hydrocarbons. Although petrol may be obtained from oil shales, the process is not economical; during the Second World War, however, the Scottish sources were used to provide about 25 gallons of petroleum per ton of shale.

Other argillaceous rocks include **fire-clays.** Fire-clay is a light-grey alumina-rich deposit often associated with coal, and criss-crossed with carbon rootlet remains. Its principal use is in the construction of furnaces. **China-clay** contains kaolinite, formed from the feldspar in granites, and is used in the ceramics industry. China-clay is white in colour and is found mostly in Cornwall. Clays used for brick-making are contaminated with sand and iron oxide, hence the characteristic red colour. The **alum shales** found in Yorkshire were formerly used as a source of alum for the dyeing and paper-making industries. **Fuller's earth** is another argillaceous rock whose absorbant properties are used in refining fats and oils.

Sandstone. Composed essentially of cemented grains of sand, sandstone has a granular texture and is the second most abundant of all sedimentary rock. In addition to quartz, sandstone may consist of sand-sized particles of calcite, gypsum or various iron compounds. **Arkoses** contains a high percentage of feldspar. Feldspar, which is usually derived from a granitic mass, decomposes rapidly in wet conditions. Therefore, its presence is an indication of rapid deposition or of prevailing desert conditions at the time of formation.

Grits. These also contain some feldspar, and were usually formed in ancient deltas. The millstone grit of the Carboniferous system is a typical example of this kind of arenaceous rock.

Greywackes are composed of ill-sorted, angular fragments including feldspar and have a fine clayey matrix. These characteristics are all indications that they were formed from neighbouring masses, rapidly deposited and cemented together, usually in a geosyncline (see page 70). Great thicknesses of this rock are found in northern Wales and Scotland.

Quartzites may be metamorphic or sedimentary; the former are termed **metaquartzites** and the latter are called **orthoquartzites.** As their name suggests, they are predominantly (or entirely) composed of quartz. These rocks are very durable and usually form prominent features where they occur, such as the Lickey Hills near Birmingham. **Ganister** is another almost pure quartz rock containing rootlets and is associated with coal-bearing sequences.

Before we leave sandstones, the **greensands** are worthy of mention. Found in the Cretaceous system in southern England, they are usually more orange than green in colour, and powder very easily. The name 'greensand' is due to the green colour of **glauconite,** an iron compound which is only stable in marine conditions (this provides a clear indication of the greensands' mode of deposition). When consolidated and exposed, the glauconite alters to another iron mineral, **limonite,** which stains the whole rock its familiar orangey-brown colour.

Conglomerate. A conglomerate may be composed of rounded pebbles of many different sizes. It is essentially gravel which has been mixed with sand and held together by natural cement. The rock fragments forming a conglomerate may range from silt-sized particles to rocks the size of boulders. **Breccias** are conglomerates composed of angular fragments, again an indication of very little transportation. Breccias may be formed from scree slopes or by crushing during faulting. **Tillites** are conglomerates formed by glacial action.

Two particular conglomerates are: the Hertfordshire Puddingstone, a Tertiary conglomerate which contains round flint pebbles; and the **Pebble Beds,** found in the Bunter Sandstone of the Triassic system, which contains pebbles from $\frac{1}{4}$ inch to 18 inch in size, from rocks of Cambrian, Ordivican and Carboniferous origin.

(2) Chemical Sedimentary Rocks

Sediments which have been precipitated from material dissolved in water are called chemical sediments. Some chemical sediments are deposited directly from the water in which the material is dissolved; for example, rock-salt may be precipitated from solution when sea water evaporates. Such deposits are generally referred to as **inorganic** chemical sediments. Chemical sediments which have been deposited by or with the assistance of plants or animals are said to be **organic, biochemical** or **bioclastic** sediments. Oysters, for example, extract calcium carbonate from sea water and use it to build a calcareous or limy shell. When the oyster dies, its shell commonly remains on the sea floor where it is eventually incorporated into the bottom sediments. Some of the commonest and most interesting chemical sedimentary rocks are listed below.

Limestone. Limestone is composed primarily of calcite, the common form of calcium carbonate ($CaCO_3$). There are many varieties of limestone with different origins and forms.

Chalk is one of the best known of British rocks. To the Englishman chalk means the white cliffs of Dover, Beachy Head and the Downs. It is typically white and often contains flints, but in Yorkshire we find red chalk, the red being due to staining by iron compounds. The formation of chalk is confined to the Cretaceous period and it occurs in countries other than Britain; it underlies most of Denmark and is also found in Australia. Its origin is still a matter of controversy. Under the microscope it can be seen to be made up of minute shell fragments and powder. It has been compared with the oozes found off Florida and this has indicated that it might be a chemical deposit. A second theory is that it was deposited in shallow seas with fine material coming in from neighbouring deserts. It is probable, however, that neither idea is completely true.

Coquina is a type of limestone composed entirely of shell fragments.

Crinoidal Limestone is composed of broken pieces of crinoids (sea lilies), creatures which typically inhabited warm shallow seas. Much crinoidal limestone is of Carboniferous age and is found in Derbyshire, Yorkshire and various other places.

Reef Limestones or *Coral Limestones*, as their name indicates, are basically composed of corals and other related coelenterates such as **Stromatopora.** They are found mainly in the Silurian (Wenlock), Devonian, Carboniferous and Jurassic systems.

Oolite and *Pisolite* are limestones composed of rounded grains precipitated around an organic nucleus, and rolled in gentle currents. (The name oolite is derived from the Greek for 'fish-roe stone', and pisolite is derived from 'pea stone'.) Often found associated with corals, the two best examples are the Great Oolite and the Inferior Oolite of Jurassic age which outcrop in the Cotswold Hills, at Bath and in Yorkshire. Today both the above rocks are seen in formation where there are coral reefs, such as the Great Barrier Reef off Australia.

Travertine forms the stalactites and stalagmites of the caves in such regions as Cheddar and Castleton. It is a crystalline, banded variety, often stained by impurities to give unusual and beautifully coloured formations.

Tufa is a spongy, porous, inorganic limestone deposited around springs and streams.

Lithographic Limestone. This is not found in Britain, but is of interest because of its fossil content. Lithographic limestone is a Jurassic rock occurring in Bavaria (southern Germany). It was formed from calcite mud, a very fine deposit in lagoonal conditions, which preserved over 400 different species, including impressions of jelly-fish and *Archaeopteryx*, a prehistoric bird, the first form to have feathers, which still had many reptilian characteristics. At one time this form of limestone was of immense commercial importance in the process of lithographic printing (the name means 'stone drawing'). The process is used today, but the stones have been replaced by metal plates.

Dolomite. Known also as magnesium limestone ($MgCO_3.CaCO_3$) dolomite is formed when some of the calcium in limestone is replaced by magnesium. The exact method of replacement is not fully understood, but reefs, oolites and fossils are completely altered by the process. Both dolomite and limestone are used extensively in the building trade.

Evaporites. These are sedimentary rocks which are derived from minerals precipitated from sea water. They include gypsum ($CaSO_4.2H_2O$), anhydrite ($CaSO_4$), and halite (rock-salt, $NaCl$). Extensive deposits of these lie beneath much of central and northern England. (More details of these are given in Chapter Two).

Coal. Coal is largely composed of carbonized plant remains and is usually found in layers, associated with other sedimentary rocks. In its initial formation coal passes through several stages: (i) *peat*, composed of partially carbonized plant material, marks the first stage, and is found in many moorland regions (central Ireland for example); (ii) *lignite* or *brown coal* occurs as a second stage. Further changes convert lignite into *bituminous* (or soft) coal, then this is altered to *anthracite* or hard coal. Coal has been, and still is, an important fuel of industry and vital to the economic prosperity of Britain. The principal coal fields are of Carboniferous age and are situated in the Midland Valley of Scotland, Northern England, the English Midlands and South Wales. The coal fields also extend into Kent.

Other organic sedimentary rocks are **radiolarite,** composed largely of siliceous exoskeletons of tiny one-celled animals called radiolarians, and **diatomite,** formed from siliceous remains of microscopic plants called diatoms. The **ironstones,** composed of various iron ores which supply much of the world's iron, are also sedimentary rocks.

(3) Physical Characteristics of Sedimentary Rocks

Sedimentary rocks possess definite physical characteristics and display certain features which make them readily distinguishable

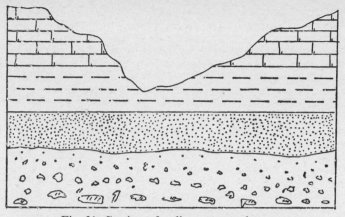

Fig. 31. Section of sedimentary rock strata

from igneous or metamorphic rocks. Some of these more important characteristics are:

Stratification. Probably the most characteristic feature of sedimentary rocks is their tendency to occur in strata or beds (Fig. 31). Each stratum or bed is separated by a **bedding plane** which marks the top of one bed and the bottom of the bed next above it. These strata are formed as geologic agents such as wind, water or ice gradually deposit their load of sediment. Changes in the carrying agent (such as reduced stream or wind velocities) will affect the texture of the sediments and the thickness of the beds.

Texture. The size, shape, and arrangements of the materials which form a sedimentary rock will determine its texture. Conglomerates exhibit a coarse texture, whereas a fine-grained limestone has a fine texture. Sands are classified as fine-grained, coarse-grained, etc. In general, textures are usually said to be *clastic* if formed from broken rock and mineral fragments, and *non-clastic* if they are more or less crystalline or granular.

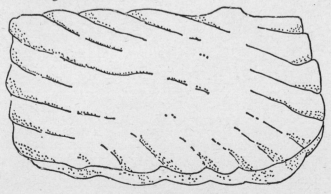

Fig. 32. Ripple marks

Ripple Marks. Little waves or ripples of sand are commonly developed on the surface of a beach, sand dune, or the bottom of a stream. Ripples of this type have also been preserved in certain sedimentary rocks (Fig. 32), and may provide the geologist with information about the conditions under which the sediment was originally deposited.

Mud Cracks. It is not uncommon to find mud cracks that have formed on the bottom of dried-up lakes, ponds, or stream beds. These polygonal (many-sided) shapes give a honeycomb appearance to the surface (Fig. 33). Such cracks preserved in sedimentary rocks

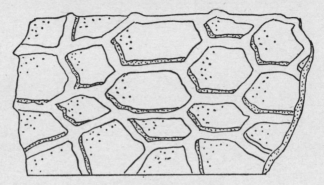

Fig. 33. Mud Cracks

suggest that the rock was subjected to alternate periods of flooding and drying. Mud cracks can be seen in the making when clay is exposed to the atmosphere during a dry spell of weather.

Various other features are apparent in sedimentary rocks. **Cross-bedding,** when present, is an indication either of the action of strong currents of water or of winds in desert regions. **Foreset beds** occur in deltas (Fig. 34) as the material brought down by fast-moving rivers

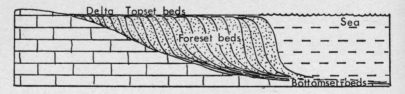

Fig. 34. Deltaic deposits

is dumped and forms the pattern shown in the diagram. These may later be truncated by erosion. **Graded bedding** occurs when material sorts itself out after quick deposition with the heavy material at the bottom. **Slumped bedding** is a preserved disturbance on the sea floor. **Varved beds** occur in association with glaciers, each layer representing the sediments released by the melting ice each summer.

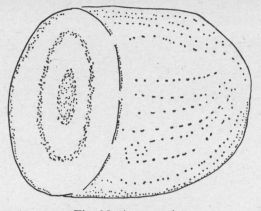

Fig. 35. A concretion

Concretions. Some shales, limestones and sandstones contain spherical or flattened masses of rock that are generally harder than the rock enclosing them. These objects, called concretions (Fig. 35), are commonly formed around a fossil or some other nucleus. Concretions range from as little as one inch to several feet in length or diameter. Because concretions are usually harder than the enclosing rock, they are often left behind after the surrounding rock has been eroded away.

Geodes. Geodes are rounded concretionary rocks that are hollow and frequently lined with crystals (Fig. 36). Geodes are most likely to occur in limestone, but they may also be found in certain shale formations.

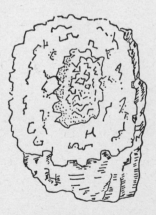

Fig. 36. Geode with crystals lining interior cavity

Colour. Such areas as Alum Bay in the Isle of Wight are noted for their brilliantly coloured formations of sedimentary rocks. The colour of these, and other rocks, is largely dependent upon the chemical composition of mineral salts which impregnate the rock. Haematite (Fe_2O_3), one of the most common colouring agents in sedimentary rocks, produces a pink or red colour. Limonite may produce yellow-coloured rocks, manganese various shades of purple, and rocks of high organic content (such as carbonaceous shale) are likely to be grey to black in colour. In addition, weathering may affect the colour of a rock. Thus, a rock that contains iron may be dark on a fresh surface, yet oxidation, brought about by chemical weathering, may result in a yellowish brown colour on the weathered surface.

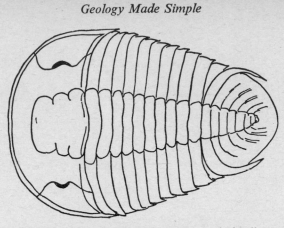

Fig. 37. Trilobite, a typical palaeozoic fossil

Fossils. Fossils are the remains or evidence of ancient plants and animals that have been preserved in the earth's crust. Fossils normally represent the preservable hard parts of some prehistoric organism that once lived in the area in which the remains were collected (Fig. 37). Only a small percentage of the plants and animals of the geologic past have left any record of their existence, and of these the majority are from a marine environment.

CHAPTER FIVE

METAMORPHISM AND CRUSTAL DEFORMATION

Metamorphic rocks are rocks (originally either igneous or sedimentary) that have been buried deep within the earth and subjected to high temperatures and pressures. Such physical conditions usually produce great changes in the solid rock, and these changes are included under the term **metamorphism** (Greek *meta*, 'change', and *morphe*, 'form' or 'shape').

During the process of metamorphism the original rock undergoes physical and chemical alterations which may greatly modify its texture, colour, structure and even chemical composition. Thus, limestone may be metamorphosed into marble, and sandstone into quartzite. Let us now consider the more important forces that can bring about metamorphic changes.

(1) Contact Metamorphism

When **country rock** (the sedimentary rock surrounding an igneous intrusion) is invaded by an igneous body it generally undergoes profound change. Hence, limestone intruded by a hot magma may be altered for a distance ranging from a few inches to as much as several miles from the line of contact between the two kinds of rock. Some of the more simple metamorphic rocks have been formed in this so-called **baked zone** of the altered country rock (Fig. 38).

Physical change may be produced by contact metamorphism when the original minerals in the country rock are permeated by magmatic fluids which often bring about recrystallization. This process, which typically produces either new or larger mineral crystals, may greatly alter the texture of the rock. In addition, the magmatic fluids

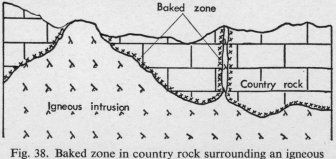

Fig. 38. Baked zone in country rock surrounding an igneous intrusion

commonly introduce new elements and compounds which will modify the chemical composition of the original rock and result in the formation of new minerals. The areas of contact called aureoles (haloes) which surround igneous intrusive rock generally form *hornfels*, a coarse-grained rock, near the intrusions, and grade into 'spotty' rocks containing such minerals as biotite and cordierite.

(2) Dynamic (or Kinetic) Metamorphism

Dynamic metamorphism occurs when rock layers undergo strong structural deformation during the formation of mountain ranges. The great pressures exerted as the rock layers are folded, fractured, and crumpled generally produce widespread and complex metamorphic change. Such pressures may result in tearing or crushing of the rocks, obliteration of any indication of fossils or stratification, realignment of mineral grains, and increased hardness. Because this type of metamorphism takes place on a relatively large scale it is also called **regional metamorphism.**

(3) Effects and Products of Metamorphism

The effects of metamorphism are controlled to a large extent by the chemical and physical characteristics of the original rock, also by the agent and degree of metamorphism involved. The more basic changes are in the texture and chemical composition of the rock. The rearrangement of mineral crystals during metamorphism results in two basic types of rock texture: foliated and non-foliated.

Foliated Metamorphic Rocks. Foliated rocks are metamorphic rocks in which the minerals have been flattened, drawn out and arranged in parallel layers or bands (Fig. 39). There are three basic types of foliation: (*a*) slaty, (*b*) schistose and (*c*) gneissic. Each of these, and some common rocks which exhibit them, is discussed below.

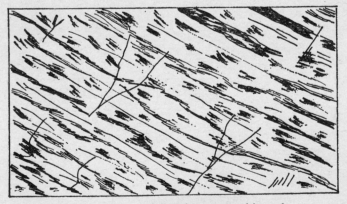

Fig. 39. Schist, a foliated metamorphic rock

(*a*) *Slate*. A metamorphosed shale, slate is characterized by a very fine texture in which mineral crystals cannot be detected with the naked eye. It does not show banding and splits readily into thin, even slabs. Slate occurs in a variety of colours, but is usually grey, black, green or red. Its characteristic slaty cleavage (not to be confused with mineral cleavage) makes it especially useful for roofing, blackboards and paving.

(*b*) *Schist*. Schist is a medium- to coarse-grained foliated metamorphic rock formed under greater pressures than those which form slate. It consists principally of micaceous minerals in a nearly parallel arrangement called 'schistosity'. Schists usually split readily along these schistose laminations or *folia*, which are usually bent

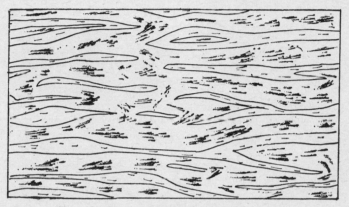

Fig. 40. Gneiss, a banded metamorphic rock

and crumpled. Commonly derived from slate, schists may also be formed from fine-grained igneous rocks. They are named according to the predominant mineral, such as mica schists, chlorite schists, and so on.

Garnet mica schists contain the silicate mineral known as garnet. The best garnet crystals, ruby red and transparent, are semi-precious stones; inferior crystals are used in industry as abrasives.

Phyllite. Derived from the Greek word *phyllon* (a leaf), phyllites are more fine-grained than schists but coarser than slate. On freshly broken surfaces they have a characteristic silky lustre, or sheen, due to the presence of fine grains of mica. Most have been formed from shales which have been subjected to pressures greater than those required to produce slate, but not of sufficient intensity to produce schists.

(*c*) *Gneiss*. Gneiss (pronounced 'nice') is a very highly metamorphosed coarse-grained banded rock. This rock is characterized by alternating bands of darker minerals such as chlorite, biotite mica, or graphite (Fig. 40). The bands are typically folded and contorted, and although some gneisses resemble schists, they do not split nearly as easily. Banding may be an indication of stratification in the original bedded sedimentary rock, or it may be caused by the alteration of coarse-grained igneous rocks containing light and dark coloured minerals.

In general, gneisses have undergone a greater degree of metamorphism than have schistose rocks, and are commonly formed as a result of intense regional metamorphism.

Non-foliated Metamorphic Rocks. These are metamorphic rocks which are typically massive or granular in texture and do not exhibit foliation. Although some non-foliated rocks resemble certain igneous rocks, they can be distinguished from them on the basis of mineral composition.

Quartzite. Quartzite is formed from metamorphosed quartz sandstone. One of the most resistant of all rocks, quartzite is composed of a crystalline mass of tightly cemented sand grains. When formed from pure quartz sand, quartzite is white; however, the presence of impurities may stain the rock red, yellow or brown. These metamorphic quartzite rocks are sometimes called *metaquartzites* to distinguish them from the sedimentary *ortho*quartzites (see page 58).

TABLE 3. Some common igneous and sedimentary rocks and their metamorphic equivalents.

ORIGINAL ROCK	METAMORPHIC ROCK
Sedimentary	
Sandstone	Quartzite
Shale	Slate, phyllite, schist
Limestone	Marble
Bituminous coal	Anthracite coal, graphite
Igneous	
Granitic textured igneous rocks	Gneiss
Compact textured igneous rocks	Schist

Marble. A relatively coarse-grained, crystalline, calcareous rock, marble is a metamorphosed limestone or dolomite. It is formed by recrystallization, and any evidence of fossils or stratification is usually destroyed during the process of alteration. White when pure, the presence of impurities may impart a wide range of colours to marble.

Anthracite. When bituminous, or soft, coal is strongly compacted, folded, and heated, it is transformed into anthracite, or hard, coal. Because it has undergone an extreme degree of carbonization, anthracite coal has a high fixed carbon content and almost all the volatile materials have been driven off.

(4) Crustal Movements: Tectonism

The crust of the earth has undergone great structural change during past periods of earth history. Even today the earth's crust is continually being altered by three major forces: gradation, volcanism, and tectonism. Gradation and volcanism have been discussed in earlier chapters; let us now see how tectonic forces have affected the earth.

As usually considered, tectonism includes those processes which have resulted in deformation of the earth's crust. Tectonic movements normally occur slowly and imperceptibly over long periods of time. But some—for example, an earthquake—may take place suddenly and violently. In some instances the rocks will move vertically, resulting in uplift or subsidence of the land. They may also move horizontally, or laterally, as a result of compression or tension. The two major types of tectonic movements are, **epeirogeny** (vertical movements) and **orogeny** (essentially lateral movements).

Epeirogenic Movements. These are relatively slow movements accompanied by broad uplift or submergence of continents. Such movements affect relatively large areas, and typically result in tilting or warping of the land. An uplift of this type may raise wave-cut benches and sea cliffs well above sea level. Features of this sort are common along certain parts of the Pacific coast of North America. Parts of the Scandinavian coast are rising as much as three feet per century. Subsidence of the continents may also take place. Thus, continental areas sink slowly beneath the ocean and become submerged by shallow seas. Similar movements have caused the British Isles to become isolated from continental Europe. (Submergence may, of course, also be caused by a rise in sea level.)

Rock strata involved in epeirogenic movements are not usually greatly folded or faulted (fractured). As noted above, however, such strata may undergo large-scale tilting or warping.

Orogenic Movements. These are more intense than epeirogenic movements, and the rocks involved are subjected to great stress. These movements, known also as orogenies or mountain-making movements, normally affect long narrow areas and are accompanied by much folding and faulting. Igneous activity and earthquakes also commonly occur with this type of crustal disturbance. Although orogenic movements are slow, they do occur somewhat more rapidly than epeirogenic movements.

(5) Rock Structures Produced by Tectonism

Tectonic movements, whether epeirogenic or orogenic, will result in rock deformation. Under surface conditions, ordinary rocks are relatively brittle and will fracture or break when placed under great stress. Deeply buried rocks, however, are subject to such high temperatures and pressures that they become 'plastic' (i.e. pliant and supple). When subjected to prolonged stress these rocks are likely to warp or fold rather than fracture.

Warping. Warping is usually caused by raising or lowering broad areas of the earth's crust. The rock strata in such areas appear to be essentially horizontal, but close study shows that the strata are gently dipping (inclined). Warping movements are typically epeirogenic and are accompanied by little or no local folding and faulting.

Folding. Not only may rocks be tilted and warped, they may also be folded (Fig. 41). Folds, which vary greatly in complexity and size, are formed when rock strata are crumpled and buckled into a series of wavelike structures. This type of structural development is usually produced by great horizontal compressive forces and may result in a variety of different structures.

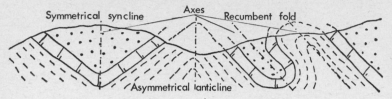

Fig. 41. Fold types

Anticlines (Fig. 41) are upfolds of rock formed when strata are folded upward. **Synclines** (also Fig. 41) may be created when rock layers are folded downward. Broad uparched folds covering large areas are called **geanticlines;** large downwarped troughs are known as **geosynclines.** Great thicknesses of sediments have accumulated in certain geosynclines of the geologic past, and some of these have been elevated to form folded mountain ranges. Such a geosyncline occupied much of Britain in the Palaeozoic. It extended from Ireland across Wales through to Scotland and Norway. The vast thicknesses of deposits which were formed in the geosyncline were later uplifted and folded to give mountainous regions.

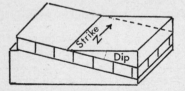

Fig. 42. Strike and dip; the beds strike north–south and dip to the east

In studying folds we must be able to determine the 'attitude' of the rock strata. **Attitude** is a term used to denote the position of a rock with respect to compass direction and a horizontal plane; it is defined by 'strike' and 'dip' (Fig. 42). The **strike** of a formation is the compass direction of the line formed by the intersection of a bedding plane with a horizontal plane. **Dip** is the angle of inclination between the bedding plane and a horizontal plane. The direction of dip is always at right-angles to the strike; thus, a rock stratum which dips due north would strike east–west.

Other types of folds include monoclines and domes. A **monocline** is a simple steplike fold which dips in only one direction. Two examples of different origin are the monocline seen in Whitecliff Bay (Isle of Wight) due to folding (Fig. 43a), and the one on the London Platform where the underlying Palaeozoic rocks have moved along a faultplane (Fig. 43b). A **dome** is a fold in which strata dip away from a common centre. The eroded dome which we know as the Weald (between the North and South Downs) is called an **anticlinorium,** being basically a huge anticline with several minor folds

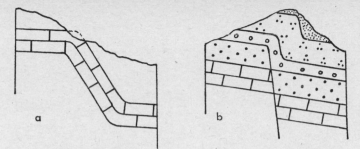

Fig. 43. Monoclines: (*a*) due to folding, (*b*) due to faulting

(the Fernhurst anticline) and faults within it. Large basins such as the Hampshire basin are similarly called **synclinoriums** (see Fig. 44). Anticlines or synclines may be **symmetrical** (i.e. the dip in each limb is the same) or **asymmetrical.** If the anticlinal fold is tilted in a secondary direction it is said to be **pitched.** When it is overfolded so that older rocks rest on younger rocks the fold is called **recumbent** (Fig. 41). Part of the Grampians is composed of a large recumbent fold.

Fig. 44. Diagrammatic representation of an anticlinorium

Fracturing. Rocks subjected to great stress near the surface are apt to fracture, thus producing joints and faults. A fracture along which there has been little or no movement is called a **joint** (Fig. 45). Joints occur in sets and are usually parallel to one another. Fractures of this sort have formed in igneous rocks as a result of contraction due to cooling, and are common in certain dykes and sills (see page 45). Joints are also created by tension and compression when rocks undergo stress due to warping, folding, and faulting.

Joint systems are developed when two or more sets of joints intersect. These intersecting joint patterns may be helpful in certain quarrying operations and in developing porosity in otherwise impervious rocks. Jointing will also hasten weathering and erosion, for they render the rocks more susceptible to attack from rain, frost and streams.

Faults. These are fractures in the earth's crust along which movements have taken place (Fig. 46). The rocks affected by faulting are displaced along the fault plane. If the crust is displaced vertically, the rocks on one side of the fault may stand higher than those on the other. This may result in a cliff called a **fault scarp.** Large-scale faulting of this type may produce 'fault-block' mountains.

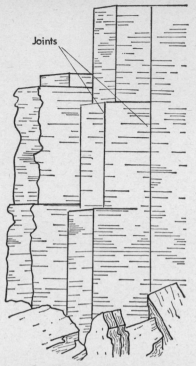

Fig. 45. Vertical joints in limestone

The terms used in describing a fault are illustrated in Fig. 46. The rock surface bounding the lower side of an inclined fault-plane is known as the **footwall,** and that above is the **hanging wall.** The **strike** of a fault is the horizontal direction of the fault-plane; **dip** is determined by measuring the inclination of the fault-plane at right-angles to the strike. More often the **hade** of a fault is given; this is the maximum inclination of the fault-plane to the vertical (and is thus

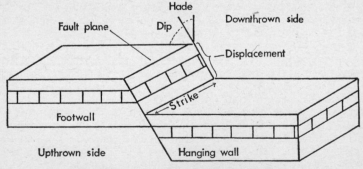

Fig. 46. A normal fault with terminology

the complement of the angle of dip.) **Displacement** refers to the amount of movement that has taken place along the fault-plane. Displacement gives rise to a **downthrown** side and an **upthrown** side. These terms are purely relative as it is not always possible to see which way the fault moved.

The various types of faults are classified largely by the direction and relative movement of the rocks along the fault-plane. A **normal** or **gravity fault** is one in which the hanging wall has moved downward with respect to the footwall (Fig. 47). If the hanging wall has

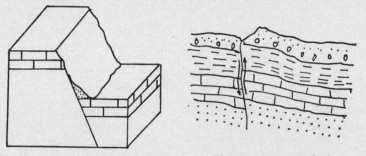

Fig. 47. A normal or gravity fault   Fig. 48. A reverse fault

moved upward with respect to the footwall, a **reverse fault** or **thrust fault** is produced (Fig. 48). A **strike-slip fault** will be produced if the movement is predominantly horizontal parallel to the fault-plane (Fig. 49).

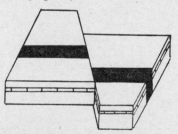
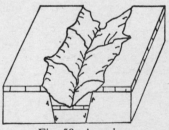

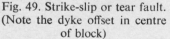

Fig. 49. Strike-slip or tear fault.   Fig. 50. A graben
(Note the dyke offset in centre
of block)

In some areas a long narrow block has dropped down between normal faults, thereby producing a **graben** (Fig. 50). Large-scale grabens are called **rift valleys.** Two examples of grabens are the upper Rhine Valley and the depression containing the Dead Sea. Sometimes blocks will be raised between normal faults; these elevated blocks are called **horsts** (Fig. 51). Examples of horsts are the Vosges in France, the Black Forest in Germany, and Charnwood Forest in Leicestershire where a block of Pre-Cambrian rock shows through the younger rocks.

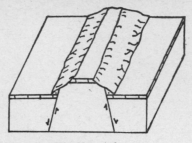

Fig. 51. A horst

(6) Evidence of Crustal Movements

The rocks of the earth's crust present much evidence to show that many tectonic movements have taken place in the geologic past. We know, for example, that the fossilized remains of sea plants and animals may be found thousands of feet above sea level. Common also are elevated beaches, coastal plains, wave-cut cliffs and sea caves. Such features strongly suggest a drop in sea level or an uplift of the continent—possibly both. Similarly, drowned river valleys indicate a rising sea and/or a subsiding land mass.

The occurrence of earthquakes is evidence that similar movements are taking place today. A good example of this can be seen in the Yakutat Bay area of Alaska. Here, in 1899, faulting caused some parts of the coast to be raised as much as 47 feet. Likewise, during the San Francisco earthquake of 1906 the horizontal movement along the fault-plane caused certain fences and roads to be offset as much as 20 feet. Britain is not in an earthquake belt, but a few years ago a minor tremor occurred in the Channel. Windows were broken in Portsmouth and slight vibrations were noticed 20 miles or so inland.

Seismographs have now also been set up on the moon to measure moonquakes.

(7) Causes of Crustal Movements

Although four theories to explain tectonic movements are outlined below, we must bear in mind that geologists do not agree on what is the exact cause in any particular instance. It may be that the movement in certain cases is the result of two or more of these causes; possibly it is the result of some other cause as yet unknown.

(*a*) **Contraction Theory.** According to this, the rocks of the outer crust have become crumpled and wrinkled as the interior of the earth has cooled and contracted. Shrinkage may also come about as great pressures squeeze the earth into a smaller volume, or when molten rock is extruded upon the surface.

(*b*) **Convection Theory.** It has been suggested that convection currents in molten rock beneath the earth's crust may cause the solid rocks near the surface to expand and push upward. The heat to produce such currents may be derived from the radioactive decay of elements such as uranium. According to this theory, circulating convection currents would exert frictional drag beneath the crust, thereby causing crustal displacement (Fig. 52).

(*c*) **Continental Drift Theory.** This theory suggests that there was originally only one huge continent. Subsequently the single continent broke into several segments which drifted apart. This 'drifting' or 'floating' was possible because the continents, composed largely of granite, were less dense than the basaltic material beneath the crust. As the front of the drifting land mass moved forward, frictional drag with sub-crustal material caused the continental margins to crumple up, thus forming the folded coastal mountain ranges of Europe and North and South America. Look at a globe and you will see how this idea originated. You will notice that the shorelines along both sides of the Atlantic Ocean match surprisingly well.

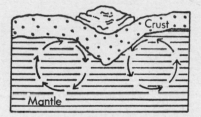

Fig. 52. Convection currents in the mantle (arrowed) and their relation to the crust

Moreover, some of the older mountain belts in America appear to be continuations of similar mountain belts in the eastern continents. Further evidence for this theory is claimed by a recent find (Autumn 1969) in Antarctica of the remains of a Jurassic amphibian which had previously only been found in America and Africa. The creature is thought to have walked into the region while the land masses were connected. If so, the separation must have taken place in or after Jurassic times. Such fossil evidence as that of the development of monkeys in various parts of the world is also used as evidence to support the continental-drift theory.

(*d*) **Isostasy.** The theory of isostasy states that, at considerable depth within the earth, different segments of the crust will be in balance with other segments of unequal thickness. The differences in height of these crustal segments is explained as the result of variations in density. Consequently, the continents and mountainous areas are high because they are composed of lighter rocks; the ocean basins are lower because they are composed of denser (heavier) rocks (Fig. 53). As the continents are eroded and sediments deposited in the ocean, the ocean basin is depressed because of the added weight of the accumulating sediments. This causes displacement of the plastic sub-crustal rocks which push the continents up. The upward displacement of the continents is aided by erosion which removes rock materials, thus making the continents lighter and more susceptible to uplift.

Because the movements of isostatic adjustment are essentially vertical in nature, this theory cannot account for forces of horizontal compression. Isostasy does, however, offer some explanation of why the erosion of the continents and subsequent deposition in the ocean

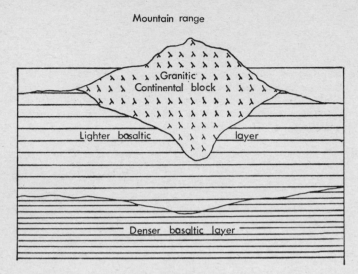

Fig. 53. Relatively light granitic rocks of continent resting on denser basaltic substratum

basins have not resulted in a continuous level surface on the face of the earth. The isostatic readjustment following the end of the Ice Age is thought to account for the rising of Scandinavia, and it is predicted that it will have to rise several hundred feet more for a balance to be obtained.

Considerable discussion has taken place on the origin of anthracites. Many authorities classify them as sedimentary rocks, but it is doubtful whether they can be placed strictly as either sedimentary or metamorphic rocks.

CHAPTER SIX

WEATHERING AND SOIL FORMATION

Weathering, the process whereby rocks undergo natural chemical and physical change at or near the surface of the earth, is one of the most important of all geologic processes. It provides much of the material from which sedimentary rocks are formed, it is important in the shaping of surface land forms, and it is responsible for the formation of soil.

Rock fragments produced by weathering are removed by erosion—the loosening and carrying away of rock debris by natural agents. Weathering and erosion are constantly at work wearing away the rocks of the earth's surface. The action of water, wind and glaciers is considered in Chapter Seven and Eight.

(1) Physical Weathering

Physical, or **mechanical,** weathering takes place when a rock is reduced to smaller fragments without undergoing a change in chemical composition. This type of weathering, known also as **disintegration,** may be the result of a variety of physical forces.

Frost Action. When water freezes in the cracks or pores of rocks it expands. The pressure exerted may be upwards of 2,000 lb. per square inch. This process of expansion may be of two kinds: **frost-wedging** or **frost-heaving.** Both kinds exert sufficient pressure to break down the rock. In frost-wedging, the pressure is directed laterally; in frost-heaving, which usually occurs in unconsolidated rocks, it is exerted upward. The latter process may force rock fragments to the surface and cause damage to posts, foundations, roads and so on. Frost action also produces screes on mountain sides, such as those around Snowdon. It also acts together with other agents; frost-action forms a scree at the base of a cliff and this is removed by the tides, forming a wave-cut platform, an example of which is the *strandflat* in Norway.

Alternate Heating and Cooling. In some areas of the earth, particularly in mountainous regions and deserts, the rocks are subjected to frequent drastic changes of temperature. Rocks on a high mountain peak expand as they are heated in the daytime, and contract when subjected to freezing temperatures at night. This process, repeated over long periods of time, will cause small cracks and crevices which permit other agents of weathering, such as frost-wedging or solution (see page 78), to attack the rock. Heat from forest and prairie fires may also hasten the physical breakdown of rocks. The process of **exfoliation,** in which curved shells of rock peel off because

77

of these temperature changes, can often be heard at night as a loud report when the rock splits. In spite of the effects outlined above, geologists are still undecided as to the precise role of temperature variations in rock disintegration.

Organic Activities. The activities of plants and animals also promote rock disintegration. Tree roots, which frequently grow in rock crevices, can be as destructive as frost-wedging. In addition, such burrowing animals as rodents, worms, and ants bring to the surface rock particles to be exposed to the action of weathering. Also included here are many of the activities of man. Large-scale rock disintegration commonly accompanies such operations as road construction, mining, quarrying, and cultivation.

(2) Chemical Weathering

Chemical weathering, or **decomposition,** produces a chemical breakdown of the rock, which may destroy the original minerals and produce new ones. Physical weathering simply produces smaller fragments of the parent rock; chemical weathering produces rock materials that are basically different from the original rock. Although chemical changes occur in a variety of ways, the more common processes of decomposition are oxidation, hydration, carbonation and solution.

Oxidation. Oxidation occurs as the oxygen in the air, assisted by water, combines with minerals to form oxides. Rocks and minerals containing iron compounds are especially susceptible to this type of decomposition. The oxidation of iron compounds, which produces rust, is also responsible for the colour of many red, yellow, and brown rocks and soils. In addition, certain compounds (for example, pyrite) form acids when oxidized. These acids attack the rocks and thereby hasten the process of decomposition.

Hydration. The chemical union of water with another substance is called hydration. Rocks and minerals subjected to this process frequently produce new compounds, especially hydrated silicates and hydroxides. Examples of this are the conversion of anhydrite to gypsum and the reaction between haematite and water to produce limonite. Hydration is also the process by which feldspars are converted into clay minerals.

In addition to the chemical effect of hydration, there is a physical expansion of the minerals during this process. This creates zones of weakness within the rock which accelerate its physical breakdown.

Carbonation. Carbon dioxide (CO_2), which is generally present in air, water and soils, commonly unites chemically with certain rock minerals, greatly altering their composition. Substances produced in this manner (carbonates and bicarbonates) are relatively soluble and therefore easily removed and carried away. In addition, the union of carbon dioxide and water produces carbonic acid (H_2CO_3), an effective agent in attacking such minerals as calcite and dolomite.

Solution. The process whereby minerals and rocks are dissolved in water plays an important part in chemical weathering. Although water has a solvent effect on all minerals, solution is greatly accelerated by the presence of carbonic acid or acids derived from the decomposition or body wastes of organisms. Most of this work is carried on by percolating ground waters which remove minerals from the rock and soil as they seep downward. Solution weathering, known also as **leaching,** may also dissolve the cementing agents in sedimentary rocks and bring about their physical breakdown. Limestones are particularly prone to this method of weathering. The jointing which is so often found in limestone corresponds to lines of weakness; caves and potholes with stalactites and stalagmites result. Wookey Hole and Cheddar Gorge, Somerset, Castleton, Derbyshire, Malham, Yorkshire (Fig. 83), and Gaping Ghyll, near Ingleborough, Yorkshire, are all examples of such weathering, often accompanied by more destructive river action which is discussed later (Chapter Seven).

(3) Rates of Weathering

The forces of erosion are continually at work removing the mantle rock, or mantle-layers of loose weathered rock materials which overlie the bedrock. Bedrock, solid rock which lies in a relatively undisturbed position beneath the mantle, deteriorates as it becomes exposed to the various agents of weathering.

The rate at which the bedrock is weathered depends largely on such factors as the composition and physical condition of the rock, climatic conditions, the kind of topography (the physical features or configuration of the land surface), and also on the structures present.

Composition of the Rock. The mineral and chemical composition of a rock is an important factor in determining the extent to which it will weather. In general, igneous rocks are relatively resistant to mechanical weathering, but more susceptible to chemical weathering. Many sedimentary rocks, especially limestones and dolomites, are highly vulnerable to decomposition by carbonation and solution. The nature of the **cementing material** (the substance holding the rock particles together) is also a determining factor. A siliceous sandstone (with silica as the cementing agent) is more enduring than a calcareous sandstone in which calcite is the cementing agent. Certain metamorphic rocks, especially quartzite, are among the most durable of all rocks.

Physical Condition of the Rock. Rocks that contain holes, cracks, and crevices allow weathering agents to penetrate deeply within them, thus hastening their destruction. Solid, unbroken rock surfaces are considerably more weather-resistant.

Climatic Conditions. Weathering, especially the chemical type, takes place more rapidly in warm, moist climates which have an abundance of rainfall. In warm, dry climates, weathering is apt to be predominantly physical in nature, and proceeds relatively slowly; the same is true of very cold regions.

Topography. Weathering proceeds most rapidly where the land slopes steeply. Rock debris is removed rapidly in such areas, and new surfaces are continually exposed to weathering processes. With an increase in altitude there is an increase in rainfall and a decrease in temperature; both of these factors will increase the rate of weathering.

Structures. More often than not an anticline is worn away, while a syncline stands out as a prominent feature. In an anticline the cap is stretched and is rapidly denuded; the reverse is often true in synclines. Examples of this **inverted topography** are seen in the Weald and in such features as the Devil's Kitchen in Snowdonia. This is a form of differential erosion mentioned below.

(4) Effects of Weathering

As bedrock is attacked by the various processes of physical and chemical weathering, some of the following special effects may be produced.

Differential Weathering. In differential weathering, different portions of an exposed rock mass weather at different rates. This type of weathering permits the more resistant parts of the rock to stand in relief after the removal of the softer or more soluble rock material. Differential weathering may come about as the result of structural or mineral differences, variations in the cementing material or the presence of concretions (see Chapter Four). Much of our scenic wealth is the result of differential weathering.

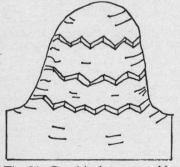

Fig. 54. Granitic dome caused by exfoliation

Exfoliation. Exfoliation occurs when thin flakes, or more or less curved slabs, peel off exposed rock surfaces (Fig. 54). There is some disagreement as to the exact cause of exfoliation, but it would appear to be the result of the combined effects of alternating temperature changes and relief of pressure. As the slabs of rock slide away they leave rounded rock masses called **exfoliation domes.** Such domes are found in various parts of the world including Africa, North America and Australia.

Spheroidal Weathering. Under certain conditions, exfoliation may continue until boulders are reduced to rounded, more or less spherical, bodies (Fig. 55). This type of weathering phenomenon is more common in fine-grained igneous rocks, but may also occur in massive shale formations.

Talus or Screes. Weathered rock debris at the bottom of a steep mountain, cliff, or slope is called talus (Fig. 56). Some accumulations of talus, known as talus deposits, or sliderock, are hundreds of feet thick. Talus deposits are generally formed as a result of frost action or some other form of physical weathering, and descend to the foot of the slope under the action of gravity.

Fig. 55. Boulder showing spheroidal weathering

Fig. 56. Talus or scree slope
formed of weathered rock debris

(5) Soils

The most important products of the forces of weathering are soils.
Soil consists of decomposed and disintegrated mantle rock which has
been altered to the extent that it can support plant life. Most soils
contain a certain amount of **humus,** dark organic material produced
by the decomposition of plant and animal materials.

Of the many factors ultimately governing the type of soil that is
developed in a given area, the more important are (*a*) composition
of the parent (original) rock, (*b*) climate, (*c*) topography, (*d*) time,
and (*e*) plant and animal activity. These factors must also be considered in the context of soil classification, discussed below.

The upper six or eight inches of the soil is called **topsoil.** Below
this is a lighter, more compact, less fertile layer called the **subsoil.**
Soils which have been derived from the bedrock upon which they
rest are called **residual soils. Transported soils** have been carried to
their present location by wind, water, gravity, or glaciers; they
typically contain rock types different from the bedrock below them.

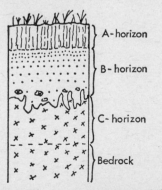

Fig. 57. Soil profiles showing A-, B-, and C- horizons

Soil Profiles. Every different kind of soil exhibits its own characteristic soil profile—a succession of **soil horizons** (or layers), normally three, each of which differs from those above or below it. The nature of these horizons in a mature, or well-developed, soil is important in soil classification and is discussed below.

The *A-horizon*, the uppermost layer in the soil profile, is the topsoil. It typically contains varying amounts of humus and has usually undergone a certain amount of leaching. Beneath this is the *B-horizon*, or subsoil, which in moist climates contains considerable amounts of clay and iron oxides but little organic material. The lowermost *C-horizon* is comprised largely of slightly altered parent rock which grades downward into the bedrock (Fig. 57).

(6) Classification of Soils

Soil scientists classify soils according to the prevailing climate in which they are developed and the vegetation which is associated with them.

Podsols are the norm for Britain and are best seen in parts of Surrey. The *A*-horizon is rich in humus, next comes a leached layer, then the *B*-horizon which is usually stained by iron oxides producing what is known as a 'larkspur'. Below occurs the bedrock, horizon *C*, which is usually partially weathered.

Brown Earths occur where forests have given rise to a thick layer of humus with little leaching. They are common in southern and eastern England.

Rendzinas occur in limestone and chalk regions. A thin, often acidic, soil rests on the chalk. This soil often bears a characteristic short grass.

Immature Soils are found in the Highlands of Britain. Much of the soil present before the Ice Age was removed by glacial action, and since then insufficient time has elapsed for the development of full soil profiles.

Pedalfers, Pedocals and **Laterites** are usually found elsewhere in the world. Pedalfers are characteristic of temperate humid climates with heavy vegetation. Pedocals rich in calcium carbonate occur in regions of low rainfall and high temperature, and usually bear grasses and brush. The last in this group, the lateritic soils, are typical of moist tropical regions with jungle vegetation.

CHAPTER SEVEN

GEOLOGIC AGENTS: WATER

Running water is undoubtedly the most important agent of erosion. It probably does more to wear away the land than all the other agents of erosion combined. Rivers are continually altering the surface features of the earth and carving out the major land forms so familiar to us all.

Each year precipitation equal to approximately 4,000 million tons of water falls on all the land surface of the earth. Although the annual rate of precipitation varies greatly from one area to another, the average annual precipitation on land is estimated to be 40 inches of water. Of this, about 22–30 per cent becomes **run-off**—water that flows from the land. Most of our surface rivers are formed as a result of this run-off.

(1) The Hydrologic Cycle

In order that we may more fully understand the origin and ultimate disposal of river waters it will help to have some knowledge of the hydrologic cycle. This cycle is a continuous process whereby water is

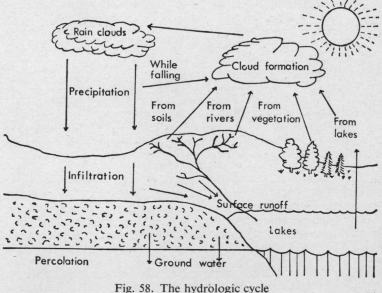

Fig. 58. The hydrologic cycle

evaporated from the seas, carried inland and precipitated as rain or snow, and eventually returned to the sea (see Fig. 58).

Most of the water on the earth has been derived from the atmosphere as rain or snow. Apart from the estimated 22–30 per cent that is carried back to the sea as run-off, much of this water sinks into the earth by infiltration to become **ground water.** Water is returned to the atmosphere by the processes of evaporation and transpiration (the exhalation of water vapour by plants). In addition to water that is disposed of by run-off, infiltration, or evaporation, some water is retained on the surface for long periods as glacier ice. (Glaciers are discussed in Chapter Eight.)

(2) Drainage Patterns and River Types

Rivers are of many different types. They range in size from the large, muddy, slowly moving Mississippi, to small, clear, rushing mountain streams. Those that flow only during periods of rain (or

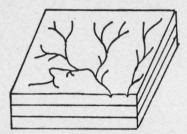
Fig. 59. Dendritic drainage pattern

Fig. 60. Radial drainage pattern

during rainy seasons) are called **intermittent streams.** Those that flow continually because they have cut their valleys to the ground-water table are called **permanent streams.**

Drainage Patterns. The two most important facts determining the course followed by a river are the slope of the land and the nature of the underlying rocks. As they make their way to the sea, rivers and their tributaries form characteristic designs or patterns. The nature of these drainage patterns aids geologists in their interpretation of the underlying rock structure and geologic history of the area.

Most drainage patterns, because of the way tributaries enter the main channel, resemble the branching of a tree and are said to be **dendritic** (Fig. 59). This treelike pattern is typical of areas underlain by flat-lying sedimentary rocks or massive igneous or metamorphic rocks. A **radial** pattern is developed when drainage channels radiate in all directions from a central area such as the top of a mountain (Fig. 60). When tributaries join larger streams at right angles, a **trellis** pattern is developed (Fig. 61). This type of pattern is typical of regions where tilted rock layers of unequal resistance outcrop. In such areas, valleys develop along outcrops of weak rock while the more resistant layers stand out as divides. A **rectangular** pattern (Fig. 62) may occur in areas where the underlying bedrock is highly fractured; this pattern is similar to the trellis pattern (compare Figs. 61 and 62).

River Types. A river may also be classified according to the relationship between its course and the underlying rocks. A **consequent** river is one whose direction of flow follows the original slope of the land. This type of river typically develops in areas of low relief, such as a coastal plain. A **subsequent** river is one whose direction of flow has been altered by fractures or folds, or by differences in the hardness of the underlying rocks. The channels of this type of river usually follow beds of soft rocks, such as shale. On the other hand, an **antecedent** river will adhere to its original course in spite

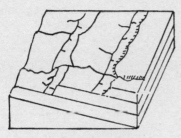 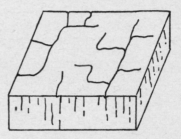

Fig. 61. Trellis drainage   Fig. 62. Rectangular drainage
      pattern   pattern

of any uplift that might have taken place along its course. (Subsequent rivers commonly show up as tributaries to the antecedent river in a trellis type of drainage pattern—see Fig. 61). **Superimposed** (or **superposed**) rivers are those that have cut through the softer overlying strata in which they originated and imposed themselves on the older underlying rocks. The rocks which underlie superimposed rivers have a different composition and structure from those in which the original channel was developed.

(3) The Work of Rivers

In general, the work of running water starts with rain, much of which is disposed of by run-off. The water may start its work as a sheet of water (**sheet-wash**), but ordinarily will soon be diverted into a river. The amount of run-off in an area may be increased by (*a*) steep slope, (*b*) low permeability of surface materials, (*c*) lack of vegetation, and (*d*) short, heavy, or prolonged gentle rains.

(4) River Erosion

River erosion occurs by means of several processes, most of which act in co-operation with each other. A variety of factors ultimately determine the degree and type of erosion that will take place in a given area, for example:

Abrasion or Corrasion. The abrasive or corrasive ability of a river depends upon its **load**—the amount of material carried by a river at a given time. Each sand grain, pebble, or boulder carried becomes a cutting tool capable of deepening and widening the river bed. Known also as **mechanical** river erosion, abrasion occurs when rock particles rub against each other or against bedrock as they are transported by the river. In addition, much damage is done by impaction when rock fragments knock against bedrock or each other.

Abrasion is most effective where river flow (velocity) is rapid, river load is heavy, and much debris is being rolled along the bottom. The effects of this type of erosion may be seen in smooth well-rounded pebbles and boulders, and in undercut banks along the sides of rivers.

Solution or Corrosion. Running water also has a corrosive or solvent effect on the rocks over which it flows. This type of river erosion occurs as water containing carbonic acid (derived from the air and vegetation) dissolves minerals in the bedrock. Such relatively soluble rocks as limestone, gypsum and dolomite are most likely to be affected by the solvent action of river water.

Hydraulic Plucking or Quarrying. The quarrying effect of a river is most effective in areas where the bedrock is highly fractured and/or poorly cemented. Strong currents may force water into bracks along the river channel, thus removing rock material from the canks or bed of the river.

Attrition. Boulders and fragments are themselves worn by friction and broken as they are rolled or carried along. The finer products of this process are then carried away with greater ease.

(5) Rate of Erosion

The rate at which running water will erode depends on the following factors:

Stream Size. Larger volumes of water are capable of carrying larger loads, which in turn increase the ability of the stream to abrade. Thus erosion will be most active when rivers are at flood stage and transporting large amounts of material.

Gradient and Velocity. Gradient refers to the slope down which the river flows. River gradients are usually higher at the source or head and become relatively low at the mouth of the river. The velocity of a river increases when gradients are steep, when there are large volumes of water, and when the channel is straight, narrow and relatively free of obstacles.

Nature of the Load. Rivers whose beds follow a straight and narrow course erode more effectively than rivers with meandering channels. Likewise, rivers whose beds contain many obstacles such as plants, boulders, and similar obstructions will be reduced in velocity and carrying power.

(6) Effects of River Erosion

The erosional work of rivers is responsible for a variety of interesting geologic phenomena. Among these are stream-cut valleys and gullies, waterfalls and rapids, pot-holes, river capture, meanders, and ox-bow lakes.

Stream-cut Valleys and Gullies. Running water moving over the earth's surface cuts a depression or channel which may ultimately become a valley. Most valleys start as gullies, which become wider, deeper, and longer with each rain. The gulley is lengthened by 'headward' erosion (in the opposite direction to the flow of water) at the point where water pours into its upper end. Continued erosion produces V-shaped valleys which are deepened by erosion of the river bed and widened by erosion of its banks (Fig. 63). Such erosion may continue until the stream reaches

Fig. 63. Typical V-shaped valley produced by river erosion

base-level—the level at which there is no gradient and the flow ceases.

The degree to which valley development will proceed depends largely upon (*a*) the volume of water flowing, (*b*) its velocity, (*c*) the nature of its load, and (*d*) the resistance of the bedrock. Some valleys are deepened more rapidly than they are widened. Such valleys may be referred to as ravines, gorges or canyons. The development of stream-cut valleys is discussed in the section on *Cycles of Erosion*, page 91.

Fig. 64. Cross-section of a waterfall. (Note the resistant limestone which caps the fall)

Rapids and Waterfalls. Where there is a sudden increase in the gradient of the river, the movement of the water is accelerated and rapids may be formed. When the river bed makes a sudden vertical or nearly vertical drop, a waterfall occurs (Fig. 64). Many waterfalls, for example, Niagara Falls, are formed as the stream bed passes from resistant rock to rock that is relatively soft. As the Niagara River plunges over a cliff capped by resistant limestone, it falls on rock material which is much less resistant. The force of the cascading water scours into the shale at the base of the cliff, undercutting the limestone and allowing it to break off. This process of undermining brings about the retreat of the falls upstream. As a result of undermining, Niagara Falls appears to be retreating at the rate of four to five feet per year. The biggest fall in England, Hardrow Force in Wensleydale (Yorkshire), is of this type. It has worked its way back through a quarter-mile gorge in a succession of Yoredale beds. A walk from the public house near Hawes takes one right behind the waterfall where the soft shales have worn back, undermining the harder rocks above.

Some falls are formed when rivers pass over igneous intrusions which are often harder, and therefore more resistant, than the country rock. The Whin Sill (see page 45) gives rise to many such falls, two of which are Cauldron Snout and High Force on the Tees. The Victoria Falls in Rhodesia rushes over a great plateau of basalt to drop nearly 400 feet into the bottom of the gorge. Others, like those of Yosemite National Park, the classic example, are formed when hanging valleys (scc Chapter Eight) enter the main valley. Visitors to Switzerland may be familiar with Europe's best example, which is found between Interlaken and the Jungfrau. Smaller examples occur in Britain in the glaciated regions of Snowdonia, the Lake District and western Scotland. Rapids, cataracts and cascades are often the remnants of old waterfalls which have retreated and become less and less high.

Pot-holes. When rapidly moving streams produce whirlpools, the swirling current drives water (and its load of sand, gravel, etc.) in a rotary motion. The rock fragments caught up in this eddy will eventually grind out a shallow, circular depression in the bed of the river. Holes formed in this manner are called pot-holes and range from a few inches to as much as 20 feet in diameter.

River Capture. Sometimes a stream will be so lengthened by headward erosion that it will intercept the banks of another stream and divert it into its own channel. This phenomenon, known as river capture, comes about as a result of differences in the rate of stream erosion. The Ouse in Yorkshire has captured many rivers. River capture is also a feature of the drainage pattern in the Weald; for example, the Rother running along the Sandgate beds is captured by the Arun and enters the sea at Littlehampton.

Meanders and Ox-bow Lakes. When the gradient of a stream is such that there is approximate balance between the amount of material that it erodes and the amount it deposits, we have what is called a graded stream. Such a stream no longer cuts downward but uses most of its energy to carry its load. However, lateral erosion or side-cutting does occur, and the valley bottom continues to grow wider. As the valley grows, the stream channel normally occupies only a relatively small part of the valley floor. This permits the stream to take a wandering, winding course characterized by many wide S-shaped bends called meanders (Fig. 65a).

Some meanders become so curved that only a narrow neck of land separates the ends of the meander. During times of flood a stream may take a short-cut across the neck, and thus isolate the meander from the rest of the stream. If water remains in the abandoned section of the channel, it is called an ox-bow lake (Fig. 65b). A fine example of a meandering river can be seen near Newhaven, Sussex. The River Jordan is also an example of a slow-moving meandering river.

Braided Streams. A braided stream is one that is characterized by a series of complex channels which unite and diverge again and again. It is also filled by numerous sand bars—accumulations of sand deposited by the overload stream.

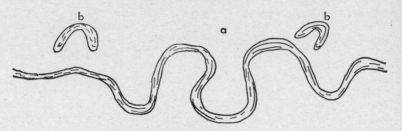

Fig. 65. Meanders: (*a*) ox-bow, (*b*) ox-bow lakes

(7) Transportational Work of Rivers

A river, like all agents of erosion, picks up most of its load as it wears away the earth materials with which it comes in contact. Each year as much as a thousand million tons of sediments are worn from the earth's surface, transported by streams, and deposited in the sea. These rock fragments will be the sedimentary rocks of future geologic time.

The carrying ability of a river depends largely on the velocity and volume of the water. But because any increase in volume is commonly accompanied by an increase in velocity, the carrying power of the river is very susceptible indeed to changes in velocity. Hence streams have greater **capacity** (carry a larger amount of material) and greater **competency** (carry larger-sized particles) during periods of flooding. Large boulders and other debris can be seen in the bed of the peacefully flowing River Lyn between Watersmeet and Lynmouth, dumped there as a result of the disastrous spates of flooding which have occurred in the past. Rock materials thus picked up may be transported in solution or suspension, or rolled along the bottom of the stream.

Dissolved Load. The dissolved materials are carried in solution. This, the so-called **invisible load,** will vary according to the degree of solubility of the bedrock with which the stream may come in contact.

Suspended Load. Most of the material transported by a stream is literally suspended or 'hung' between the bottom and the surface of the stream. Typical sediments transported in this manner are sand, silt and clay.

Bed Load. Large quantities of rock fragments are transported by being rolled or slid along the river bed. In addition, rock particles may be moved by **saltation**—series of short jumps. Rock fragments transported by any of the movements described above are said to be moved by traction, and make up the 'bed load' of the river.

(8) Deposition

A river deposits its load when the competency and capacity of the river are decreased. Some of the causes of such decreases are (*a*) reduction in stream gradient, (*b*) decrease in volume, (*c*) loss of velocity, (*d*) obstacles in the stream channel, (*e*) widening of the stream bed, (*f*) overloading, (*g*) freezing, and (*h*) emptying into quiet or slower-moving bodies of water. Material thus deposited is called **alluvium.**

Alluvial deposits contain materials which have been sorted according to size, and as a consequence are **stratified** (deposited in layers) with the coarser materials on the bottom. In addition, alluvial materials are usually composed of rock fragments which have been smoothed or rounded by stream abrasion.

Alluvial Fans and Alluvial Cones. Alluvial fans are moderately sloping fan-shaped deposits found at the foot of mountains. Characteristic of semi-arid zones, these accumulations of silt, sand, gravel, and boulders are deposited when fast-flowing mountain rivers lose their gradient and flow out on level ground at the foot of the mountain. Deposits of this type which have very steep slopes are called alluvial *cones.* Screes, in which gravity is responsible for the fragments falling, usually rest at about 40° to the horizontal, whereas in alluvial fans the angle of rest is considerably less. In certain areas of the western United States a number of alluvial fans join each other at the foot of a mountain range to form **piedmont alluvial plains.**

Deltas. When a river flows into a large body of water such as a sea or lake, its velocity is suddenly decreased and much of its load is dropped. Deposits formed under these conditions are called deltas. Some deltas, such as those of the Mississippi River (covering an area of approximately 12,000 square miles) and the Nile (10,000 square miles), are extensively cultivated because of the fertility of the alluvial deposits.

As the delta becomes larger, the main stream may overflow and form new channels called **distributaries.** Deltas are not common in Britain, but can be seen where rivers run into lakes, as in the Lake District. In the past, deltas have played a prominent part in the shaping of the geologic history and structure in Britain, as in the junction of the marine Devonian and Old Red Sandstones in south-west Britain. Deltas are still common, however, in other parts of the Commonwealth such as India, Pakistan and Nigeria.

Flood Plains. Known also as **river plains** or **valley flats,** flood plains are formed when a river in flood overflows its banks. The velocity is reduced when the stream leaves its channel and much of the load is deposited on the valley floor. Many rivers in Britain develop flood plains along their lower reaches; in exceptional years the flooding can be very serious.

River Terraces. The remnants of a flood plain that has undergone erosion are known as river terraces. Such terraces are topographically higher than the surrounding flood plain. Terraces at distinct levels are found above the Thames flood plain.

Natural Levees. When a stream flows over its banks on to the flood plain, the greatest loss of load takes place along the banks of the channel. The coarsest materials of the load are deposited here, and this process tends to build up a ridge or embankment called a natural levee. Some of these may rise 20 feet above the surrounding flood plain and afford protection to the adjacent lowlands during time of flood.

(9) Cycles of Erosion

According to some geologists, erosion follows a rather well-defined cycle, although we seldom find a perfect example of a completed cycle. Sometimes a cycle will be interrupted by **rejuvenation.** This occurs when an area that has been eroded almost to base level is again elevated, bringing about increased river gradients and accelerated erosion. In general, however, the erosion cycle begins with the erosion of an area right down to base level, followed by a fresh uplift, when erosion begins anew.

Although geologists disagree as to the validity of this concept, there is no doubt that it does provide us with much information about the evolution of landscapes. How this concept has been applied to the development of river valleys and regional land forms is discussed below.

(10) River-valley Erosion Cycle

When tectonic movements lift a region well above base level, the rivers proceed to erode the surface back down to base level. The end result of this continued erosion is a **peneplain** (or **peneplane**), an

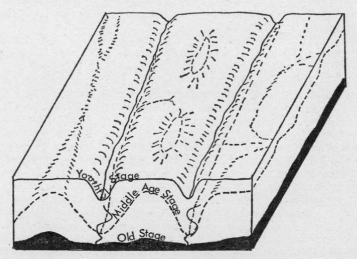

Fig. 66. Erosion cycle

extensive area of low relief and low elevation. Such peneplained levels are apparent in the erosion of the Weald Anticlinorium; there is, for example, a pronounced levelling off at about 200 feet above sea level. This erosional process, known as the river-valley erosion cycle, occurs in three well-defined stages known as youth, maturity, and old age. (The same stages are experienced by the rivers which occupy these valleys.) The character and features of the stages of the cycle of river valley erosion are shown in Table 4 and Fig. 66.

Youthful Stage. Youthful valleys are deep, steep-sided, and V-shaped; no food plains have been developed. Streams occupying youthful valleys are still well above base level and are actively engaged in cutting downward into their beds. These are said to be

TABLE 4. *River-valley erosion cycle*

	YOUTH	MATURITY	OLD AGE
Transverse profile	∨	∪	∼
Gradient	Steep	Moderate	Low, gentle
Erosion	Deepening valley	Widening and deepening	Widening valley
Channel trend	Straight	Begins to meander	Broad meanders
Valley bottom	Little or no flood plain	Distinct flood plain	Broad flood plain
Tributaries	Few and small	Maximum number	Few and large
Special features	Rapids, waterfalls	Some ox-bow lakes	Many ox-bow lakes, natural levees, marshy flood plain

youthful streams and are characterized by fairly straight courses, rapids, waterfalls, and relatively few tributaries. Such features are most likely to be present near a stream's headwaters.

Mature Stage. As rivers deepen their valleys they decrease their gradients. At this stage, there will no longer be rapids or waterfalls, and the channel will normally develop meanders. Valley maturity is further indicated by flat-floored valleys with distinct flood plains, wide meander belts, and occasional ox-bow lakes.

Old-Age Stage. Continued erosion will result in valleys that are extremely broad and shallow and marked by extensive flood-plain deposits and numerous ox-bow lakes. Such valleys, and the streams responsible for them, are said to be in old age. Old streams are normally rather sluggish, with a low gradient, and their channels are marked by many meanders, ox-bows and natural levees.

Interruption of the River-valley Cycle. It was mentioned above that the cycle of valley development may be interrupted by rejuvenation. When this occurs, the streams, as a result of increased gradients, will cut deeper into the valley floor where they may form a series of step-like terraces. Rejuvenation may also permit the stream to produce entrenched, or incised, meanders. These develop when an uplifted meandering stream continues to follow its original winding pattern but cuts deeply into the underlying rock.

(11) Regional Erosional Cycle

The **uplands** (elevated areas between rivers) are also affected by an erosional cycle. However, the valley cycle will often not proceed at the same rate as the regional (upland) cycle. The two cycles must, therefore, be considered separately. As in the previously discussed cycle of valley development, the regional cycle is marked by youth, maturity, and old age (see Table 5).

TABLE 5. *The regional cycle of erosion*

	YOUTH	MATURITY	OLD AGE
Transverse profile	⌒	⌢	—
Topography	Flat upland	Hilly, mostly in slope	Flat lowland
Drainage	Poor	Good	Poor
Relief	Increasing	Maximum	Decreasing
Special features	Lakes and marshes in upland	A few ox-bow lakes in lowland	Lakes and marshes in lowland, monadnocks

Youthful Stage. The youthful stage begins with the uplift of a land mass (such as a newly uplifted flat coastal area), followed by a period of relative stability. Youthful regions have youthful streams, deep V-shaped valleys, partially dissected uplands, and considerable relief.

Mature Stage. On reaching maturity, the region will be characterized by a completely dissected, well-drained upland. In addition, topography is generally rugged, and maximum relief and drainage have been developed. River valleys are, in general, mature.

Old-Age Stage. As erosion continues, the region will pass from maturity into old age. At this stage the upland areas and rolling slopes have been reduced to a peneplain, and relief is at a minimum. The surface is marked by a few large meandering streams in broad flat valleys. These old valleys normally have extensive flood plains and well-developed natural levees (Table 5). In some areas, hills of very resistant rock rise above the surface of the peneplain. These isolated erosional remnants are called **monadnocks.**

Interruption of the Regional Erosional Cycle. Just as a change in base level interrupts the valley cycle, so will it interrupt the regional cycle. Rejuvenation may be brought about by renewed uplift and the region thus affected will be subjected to a new cycle of erosion.

The following points are pertinent to the foregoing discussion of the cycles of erosion:

(*a*) The cycles outlined above describe features as they would develop in a humid temperate climate. A somewhat different set of features might be produced under arid or arctic conditions. The erosional features of the arid regions of Australia are vastly different from those in Britain. Apart from dry creek beds there are no true rivers, and no lakes other than the so-called 'salt flat' lakes. From the air the land appears flat, although it undulates gently between 500 and 1,000 feet above sea level.

(*b*) The terms youth, maturity, and old age are applied not only to regional topography and valleys, but to rivers occupying these valleys.

(*c*) Seldom is it possible to distinguish clearly between stages in the cycle. The transition from one stage to the next is gradual, thus an area may possess features of both maturity and old age, or youth and maturity.

(*d*) Different parts of a valley or region may be in different stages of the cycle at the same time. In general, a stream has youthful features near the source and becomes more mature towards its mouth.

(12) Ground Water

Water which falls as rain may be disposed of by run-off, evaporation or infiltration. The water that seeps into (or infiltrates) the earth is called 'ground water'. Known also as sub-surface or underground water, this water is found in the pores and crevices in the rocks and soils of the upper part of the earth's crust.

Most of the water which is found in the ground is derived from rain or snow which soaked into empty spaces in the rocks. This is the most important source of ground water.

A relatively small percentage of ground water originates beneath the earth's surface and is formed chemically from igneous masses which are buried there. It is known as **magmatic** water or, because it is making its initial appearance in the hydrologic cycle, **juvenile** water.

Connate water is water that was trapped in a sedimentary deposit at the time the sediments were accumulating. It is this type of water which is commonly found associated with the deposits of petroleum. Connate water is typically salty and may be ancient sea water which was trapped in the sediments at the time of deposition. Like magmatic water, connate water forms only a small portion of the ground water.

Gravity causes ground water to percolate downward and to fill all available empty spaces from the bottom upward. This is possible because rocks that are porous will absorb much of the infiltrating surface water. If the rocks are permeable, they will permit the ground water to move freely through them, and underground circulation is further facilitated. Rocks that are porous and permeable and permit the free flow of ground water are called **aquifers.**

Rocks differ greatly in their degree of porosity and permeability. Porosity may vary from less than 1 per cent in certain massive, unweathered igneous rocks, to as much as 30 per cent in certain sandstones. Some rocks, though not originally porous, may become permeable as a result of solution cavities, fractures, etc. Rocks which will not permit water to go through them are said to be *impermeable* or *impervious.*

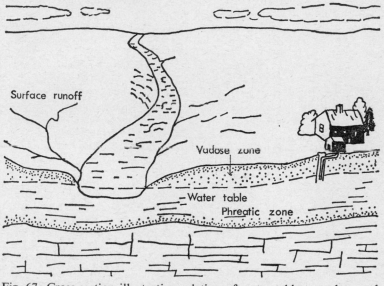

Fig. 67. Cross-section illustrating relation of water table to vadose and phreatic zones

(13) The Water Table

That portion of the earth's crust in which all available pore space has become filled with water is called the **zone of saturation** or **phreatic zone** (Fig. 67). The upper limit of this zone is known as the water table.

The rocks and soil through which ground water passes on its way to the zone of saturation are called the **vadose zone** or the **zone of aeration.** The rocks in the vadose zone never become completely saturated, for there is always some air present. This combination of air and water can act to hasten the chemical decomposition of the surrounding rocks.

The depth at which the water table will be encountered beneath the surface varies greatly in different regions. Two of the more important factors that affect the level of the water table are (i) the amount of rainfall in the region and (ii) the topography of the land.

For example, the water table may be lowered considerably during a drought, causing wells to go dry. On the other hand, a particularly wet season may result in the water table being brought nearer the surface. In general, the water table conforms roughly to the surface configuration of the land, except that it is nearer the surface in valleys and occurs at greater depth in the hills and mountains.

Wherever the water table intersects the surface of the ground, a body of water will collect to form ponds, lakes or swamps.

If ground water collects in an aquifer which overlies an impermeable layer, it becomes separated from the normal water table (Fig. 68). We then have what is called a **perched water table** because it is found higher than the normal water table.

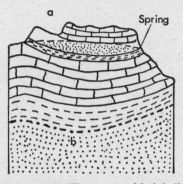

Fig. 68. A perched water table. The water-table labelled *a* is described as 'perched' because it is above the normal water table (labelled *b*)

(14) Forms of Ground Water

Although man brings much water to the surface by means of wells, most ground water reaches the surface as a result of natural seepage, as in springs.

Wells. A hole dug into the ground to a depth which reaches the water table is called a well. If it is to be considered permanent, a well must be so deep that the water table will never be below its bottom, even during very dry periods. Wells should penetrate the zone of saturation as deeply as possible, and great care must be exercised to see that they do not become polluted.

Artesian wells are wells in which the hydrostatic (water) pressure is sufficient to cause the water to rise above the level at which it was first encountered (it may or may not flow out on to the surface of the earth). Artesian wells may produce great quantities of water, and since they are not dependent on seasonal rainfall they tend to be more reliable than ordinary wells.

Certain conditions must prevail before an artesian well can be developed. The aquifer (sandstone, gravel or chalk) should dip away from the surface and have impermeable rocks above and below it

(Fig. 69). In addition, the aquifer should be exposed in an area of adequate rainfall to replenish the artesian system, and at an altitude higher than the site of the well. These conditions will produce hydrostatic pressures sufficient to develop artesian wells. The London basin has many artesian wells sunk into the chalk. One example of an artesian well can be seen from the Portsmouth train just before Petersfield station.

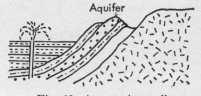

Fig. 69. An artesian well

Springs. A spring is developed when ground water comes to the surface and flows out more or less continuously. **Hillside springs** occur in hilly regions where the water table intersects the land surface (Fig. 70). **Fissure springs** are naturally formed artesian wells. The water reaches the surface through fissures (cracks) in the rocks and flows out with considerable force. **Hot springs** (or **thermal springs**) contain water that has been warmed (perhaps to boiling point) by contact with heated rocks in the sub-surface. Such areas as Hot Springs in Arkansas, Yellowstone National Park in Wyoming and Rotorua Spa in New Zealand, are famous for their hot springs.

Fig. 70. Hillside spring along bedding plane

The water in a **mineral spring** contains an unusually large amount of dissolved salts such as sodium chloride, potassium bicarbonate and magnesium sulphate, and sometimes gases such as carbon dioxide or the evil-smelling hydrogen sulphide.

Geysers are rather specialized hot springs which erupt intermittently (see page 54).

(15) Erosion by Ground Water

Ground water is an effective agent of erosion, transportation and deposition, and does its work on a relatively large scale. It erodes principally by chemical action and carries most of its load in solution.

Frequently carbon dioxide from the air and decaying organic matter in the soil combine with ground water to form carbonic acid. The acid-charged ground water makes a very effective agent of erosion, especially in areas of soluble sedimentary rocks.

As ground water passes through rocks such as limestone or dolomite, the rock is dissolved and carried away in solution. Consequently, the underlying bedrock may become honeycombed with caverns, and sinkholes and natural bridges may develop on the surface. Land surfaces characterized by such features are said to have **karst topography** (named after the Karst region of Yugoslavia). The common limestone features of Britain are sometimes referred to as karst features, but our plentiful rain and therefore abundant vegetation make them different from the true karst.

Caverns. Caverns (or **caves**) are formed when ground water, by means of solution, enlarges cracks into a series of underground tunnels and chambers (Fig. 71). The roof of the cavern is commonly formed of rock which has been able to resist the solvent action of the ground water.

Fig. 71. Caverns and sink-holes developed in an area of karst topography

Caves have long interested man and probably provided his first shelter. Peak Cavern in Castleton (Derbyshire) has provided shelter in wartime, and at one time cottages were built in its huge mouth. This cave, like others in Castleton and at Cheddar, is a tourist attraction.

Sink-holes. Occasionally caverns and other cavities are so close to the earth's surface that their roofs collapse, leaving a more or less circular depression. These holes, called sink-holes or sinks, occur commonly in regions of karst topography. Sometimes streams flow into sink-holes and disappear underground; they are then called **lost streams.** Some sink-holes become filled with debris and collect water to form ponds or lakes.

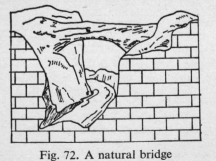

Fig. 72. A natural bridge

Natural Bridges. A natural bridge (Fig. 72) may develop when a stream flows into an opening in the rock, dissolves a tunnel, and emerges on a cliff or slope on the other side. Portions of the tunnel may then collapse and be carried away; the natural bridge is all that remains of the tunnel.

(16) **Deposition by Ground Water**

When ground water becomes oversaturated with mineral material, it is forced to deposit some of it. This deposition may be caused by variations in (*a*) temperature, (*b*) pressure, or (*c*) the loss of water by evaporation. Some of the depositional features formed by ground water are discussed below.

Spring Deposits. Certain ground waters contain such large quantities of dissolved materials that they deposit their load shortly after reaching the surface. Such deposits often form terraces and cones around hot springs and geysers. Calcareous material formed in this manner is called **travertine**; exceptionally porous travertine is referred to as a **calcareous tufa.** When siliceous material is deposited around hot springs, it is known as **siliceous sinter,** or **geyserite** when deposited around the vent of a geyser.

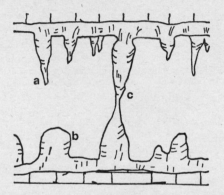

Fig. 73. Cave formations: (*a*) stalactite, (*b*) stalagmite, (*c*) column

Cavern Deposits. As might be suspected, much ground water deposition takes place in the sub-surface. Calcium carbonate in the form of travertine is most commonly deposited, but concentrations of gypsum and rock-salt may also occur.

Calcite-laden ground water dripping continuously from the same spot will eventually produce calcareous icicle-like deposits called **stalactites** (Fig. 73*a*). Some stalactites are solid, others are hollow, with the mineral-bearing water dripping through them. These hang from the cavern roof. **Stalagmites** are mound-shaped masses of calcium carbonate which are built up on the floor of the cavern at the point where water drips from a stalactite (Fig. 73*b*). Stalactites and stalagmites may eventually unite to form a **column** (Fig. 73*c*).

Cementation. Ground water plays an important part in the cementation of rock particles. This occurs when minerals carried by ground water are deposited between loose grains, thus bonding them together. In this way a loose sand may be converted into a hard sandstone when silica is deposited between the constituent sand grains.

Concretions. Concretions are formed when ground water precipitates minerals around some object such as a leaf (Fig. 74), shell or pebble. Concretions may assume a variety of shapes and sizes.

Geodes. More-or-less spherical cavities which have been wholly or partially filled with inward-pointing crystals are called geodes.

Fissure Deposits or Veins. When supersaturated ground water is deposited in cracks or crevices, mineral veins are formed. Calcite and quartz veins are common, and such veins sometimes carry concentrations of metallic minerals such as gold, silver, and copper.

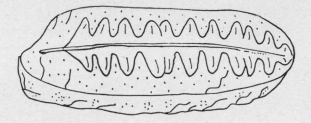

Fig. 74. Fern leaf in limestone concretion

Replacement or Petrification. The process whereby ground water dissolves one type of material and substitutes another in its place is known as 'replacement'. When organic material has been replaced we say that 'petrification' has occurred. Thus, when the woody cells of a tree are replaced by silica, petrified wood is formed (see page 173).

(17) Ground Water and Man

Ground water serves man in a variety of ways. It is vital to agriculture because some plants require large amounts of water, which they absorb from the soil through their roots. Crops in arid regions must be raised with the aid of irrigation, and much water for irrigation is derived from wells.

Industry also relies heavily on water supplies. Although much surface water is used for industrial purposes, sub-surface water must be utilized in areas where surface waters are not available.

(18) Rain

Before leaving this chapter on the action of water as a geological agent we should note that rain, continually falling on an area, loosens particles and washes them into brooks and streams. In arid regions the rare rains form gashes and ravines where the ground is underlain by clayey deposits.

Many of the features described in the last two chapters and in the next can be seen at Malham in Yorkshire, a diagram of which appears on page 110 (Fig. 83).

CHAPTER EIGHT

GEOLOGIC AGENTS: GLACIERS, WIND AND GRAVITY

Glaciers are large, slow-moving masses of land ice formed by the recrystallization of snow. These great bodies of flowing ice once occupied as much as 30 per cent of the world's land area, but today cover little more than 10 per cent.

Periods of glaciation have occurred several times during the history of the earth, but the great Ice Age of Pleistocene time (Chapter Eighteen) has provided us with the clearest and most recent record of glacial activity. During this time a great ice sheet covered much of North America and Europe (Fig. 75). At the same time the glaciers which had, since Permian days, periodically invaded Tasmania and south-eastern Australia retreated, bringing drastic changes and causing the extinction of many animal and plant forms as the region became hot and arid; rivers dried up and lakes evaporated. The last great ice sheet retreated from Britain between 10,000 and 15,000 years ago, but before it did, the ice had reached as far south as the Thames–Severn line, and south of this the hills were snow-capped.

Glacial periods normally produce great physical and biological changes in the region of glaciation. During Pleistocene time the sea level was lowered as the water was frozen into glacial ice. When the glaciers began to melt, much of the water was returned to the sea, causing the sea level to rise. In addition to bringing about changes in sea level, there is considerable evidence to indicate that the earth's crust sagged or became warped under the great weight of ice sheets. Other changes were the formation of new lakes and swamps (including the Great Lakes), alterations in the courses of rivers, and the migration of plants and animals as climates became colder.

Let us now examine the effects of glaciation, both past and present, and learn more about the role of these great ice masses in shaping the earth's surface and its history.

(1) Origin of Glaciers

In areas where snowfall is sufficiently heavy and the average annual temperature sufficiently low, the snow remains throughout the year. In regions with temperate or tropical climates, this happens on only the highest mountains. However, in the frigid zones even land which lies only slightly above sea level may be perpetually covered by snow and ice. The lower limit of perpetual snow is known as the **snow line.** Determined largely by latitude, the snow line is found at lower altitudes as the latitude increases. For example, at latitude 90°N (the North Pole) the snow line is at sea level. At latitude 0° (the equator) the snow line may be as much as 18,000 feet above sea level.

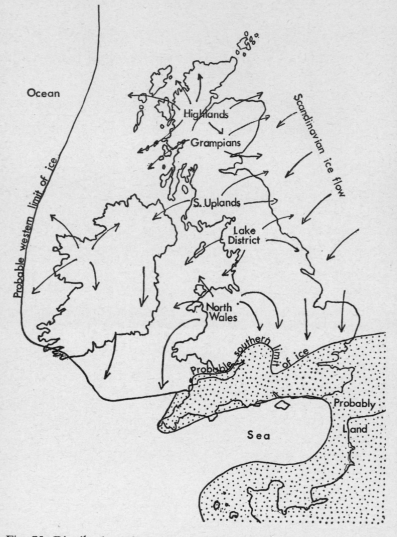

Fig. 75. Distribution of glaciers (shaded area) during Pleistocene time

It is above the snow line, in massive accumulations of snow called **snow fields,** that glaciers are born. As the snow freezes and compacts, it turns into granular, pellet-sized ice particles called **névé** or **firn.** This material, covered by subsequent snows, is gradually compressed until finally the lower level of the snow field is compacted into a great mass of ice. Eventually, the névé undergoes changes which transform the entire mass into glacier ice. When sufficient glacier ice has accumulated, the force of gravity tends to move it slowly downhill.

(2) Types of Glaciers

Geologists usually divide glaciers into three groups: (*a*) valley glaciers, (*b*) piedmont glaciers, and (*c*) ice sheets or continental glaciers.

(*a*) **Valley Glaciers.** Known also as alpine or mountain glaciers, valley glaciers originate in snow fields at the heads of mountain valleys. Their downward movement follows old stream-cut valleys, which are sometimes filled from wall to wall with glacier ice. Valley glaciers range in area from a few hundred square yards to several square miles, and in length from a few hundred yards to 75 miles or more. Such glaciers occur particularly in the Alps and in the Himalayas, also in South Island, New Zealand.

(*b*) **Piedmont Glaciers.** Sometimes two or more valley glaciers emerge from adjacent mountain valleys on to the plains below. Here the lower ends of the glaciers unite to form a broad rounded mass of ice called a piedmont glacier. One of the better-known glaciers of this type is the Malaspina Glacier on the western side of Yakutat Bay in Alaska. Covering an area of about 1,500 square miles (4,000 km²), Malaspina Glacier is formed by the union of several valley glaciers originating on the slopes of nearby Mount St. Elias.

(*c*) **Ice Sheets or Continental Glaciers.** The broad ice masses that cover large land areas are known as ice sheets or continental glaciers (smaller, relatively localized ice sheets are called **ice caps**). Usually very thick, these glaciers spread outward from the centre of a land mass until they may cover much of a continent—highlands and lowlands alike. Occasionally an isolated mountain peak known as a **nunatak** will be found projecting above the surface of the ice.

The world's largest ice sheet is that covering most of the continent of Antarctica—an area twice the size of Australia. In places this great ice mass is as much as 10,000 feet thick. The Greenland ice sheet has a surface area of approximately 670,000 square miles (1,700,000 km²) and a maximum thickness of possibly 11,000 feet.

(3) Movement of Glaciers

To the casual observer a glacier appears to be a stationary mass of ice. But glaciers, like rivers, move—although much more slowly. The rate of forward progress ranges from less than an inch to as much as 100 feet per day. Factors which affect the rate of movement include (*a*) size of the glacier (the thicker it is the faster it will move), (*b*) slope and topography of the land, (*c*) temperature of the area (the glacier moves faster as temperatures become higher), and (*d*) the amount of unfrozen water in the glacier.

The nature of glacial movement is a rather complex process. In general, however, glaciers appear to move as the force of gravity and pressure from the weight of accumulating ice causes the ice in the lower levels of the glacier to become plastic (i.e. pliant and able to flow). Ice in this lowermost portion of the glacier is said to be in the **zone of flow**; the less plastic ice in the upper part of the glacier occupies the **zone of fracture** (see Fig. 76). In addition, alternate periods of melting and refreezing produce contraction and expansion of the ice which leads to further glacial motion.

Glacial movement is somewhat like the movement of a river, in that glaciers move more rapidly in the middle than along their sides, and faster on top than along the bottom. Friction along the sides and bottom of the glacier retards movement there. As a glacier follows the twisting path of its valley and passes over irregularities in the valley floor, tension may cause fractures to develop in the rigid, more brittle ice in the zone of fracture. These fractures produce fissures, called **crevasses** (Fig. 76), some of which may be hundreds

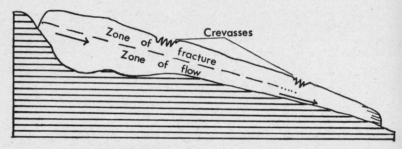

Fig. 76. Cross-section through a typical valley glacier

of feet long. Crevasses may be concealed by a thin crust of snow which breaks under the slightest weight. Because of this they are a constant source of danger to persons travelling over the glacier's surface.

Glaciers continue to move until they reach an area where warm air causes the ice to melt as fast as the glacier advances. The ice front then becomes stationary. The glacier will eventually retreat if the ice wastes more rapidly than it advances.

Many glaciers continue to move until they reach the sea, where large pieces of the glacier may break off to form **icebergs**. These float away and eventually melt as they reach warmer waters.

(4) Glacial Erosion

Like all other geologic agents, the work of glaciers consists essentially of erosion, transportation, and deposition.

So great is the erosive power of glaciers that few objects will deter them as they plough down their valleys or override prominent land features to blanket an entire continent. Glacial erosion is accomplished in one or more of the following manners:

(*a*) **Plucking** or **quarrying.** The glacier dislodges and picks up protruding fragments of bedrock.

(*b*) **Abrasion.** The ice-plucked blocks and other rock debris scratch and polish the bedrock over which the glacier passes.

(*c*) **Ploughing** or **sledding.** Loose material is pushed ahead of the glacier, or rock material may fall from valley walls and come to rest on top of the ice.

Erosion by Valley Glaciers. A valley glacier is a very effective agent of erosion and will greatly modify the area it occupies. Valley glaciers originate in **cirques** or **corries,** great semicircular depressions which have been developed by deepening and enlarging the head of

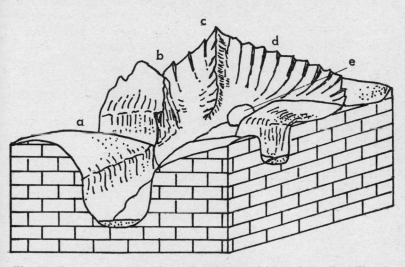

Fig. 77. Features developed by valley glaciers: (*a*) hanging valley, (*b*) col, (*c*) horn, (*d*) corrie, (*e*) tarn

a mountain valley. When several corries develop close together, as in the classic instance of Snowdon, the ridges separating them may become very sharp and jagged, producing knife-edged ridges called **arêtes.** When two corries erode into a ridge from opposite sides, they may meet to form a **col** or **pass. A matterhorn peak,** or horn, is produced in late stages of glaciation after the summits of pre-glacial mountains have been reduced to isolated peaks. When glacier ice disappears, a corrie may become filled with water to form a glacial lake called a **tarn,** for example Malham Tarn in the West Riding of Yorkshire, and many of the lakes of North Wales. Some of these features are illustrated in Fig. 77.

Valleys containing glaciers will be modified in a number of ways as the glacier grinds away the valley walls and smooths the valley floor. Young valleys will be widened and deepened and the form changed from a V to a U shape (Fig. 78), for example Llanberis Pass and Tal-y-llyn near Cader Idris in North Wales. In addition, the valley walls and floor are scratched, grooved and polished by the rock debris being carried in the glacial ice. Marks thus produced are called **glacial striae** if they are scratches, **glacial grooves** if they are very deep. Projections in the valley floor are smoothed and rounded to form **roches moutonnées** (Fig. 79), a French term meaning 'rocks resembling a fleecy wig'. Moreover, much of the valley floor over which the glacier moves may be ground to powder or **rock flour.**

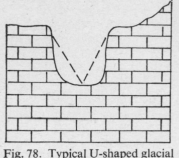

Fig. 78. Typical U-shaped glacial trough or valley

The main glacial valley (known also as a **glacial trough**) is usually more deeply eroded than are the tributary or side valleys leading into it. Thus, when the glaciers melt, these tributaries are left hanging high above the main valley floor. Valleys of this type are called **hanging valleys,** and streams flowing through them commonly form steep waterfalls where they plunge into the deeper main valleys. Yosemite Falls in California is a typical example of this kind of waterfall. Although this part of California is a classic glaciated area, Snowdonia and other parts of Britain exhibit many of the features mentioned above.

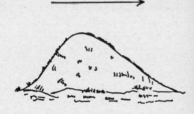

Fig. 79. A *roche moutonnée* or 'sheep rock'. Arrow shows direction of ice movement

Fig. 80. A drumlin

As a glacier flows from the mountains into the sea, the ice may excavate its valley below sea level. When the glacier melts, the sea will invade the valley, thereby producing a **fiord.** These deep, steep-walled, narrow arms of the sea are especially characteristic of the coast of Norway. Fiords are also found in South Island, New Zealand.

Erosion by Ice Sheets. Ice sheets, because of their great thickness and extent, may completely override major surface features. Thus mountains and hills as well as valleys may be subjected to glacial erosion. Beneath the ice the rocks will be quarried, scratched, and grooved, as evidenced by glacial striations and roches moutonnées. Ice sheets may also produce **drumlins,** streamlined, elliptical hills formed parallel to the direction in which the ice moved (Fig. 80). Drumlins are common in many of the Yorkshire dales.

In general, continental ice sheets tend to smooth out surface irregularities which inhibit movement of the ice sheet, and soil and mantle rock may be removed from large areas.

(5) Glacial Transportation

Glaciers are capable of carrying great quantities of earth materials, and some of these rock fragments may be quite large. Consequently, a glacier's load will normally include finely pulverized rock flour and huge boulders, with all sizes of rocks in between.

Much of this material is transported on top of the glacier and constitutes the **superglacial load.** The material frozen in the interior of the glacier is termed the **englacial load,** while the **subglacial load** consists of the rocks and soil carried on the bottom of the ice mass. It is the subglacial load that is responsible for much of the abrasive action of a glacier.

Valley glaciers carry more materials on their surface than do ice sheets or continental glaciers. This is because much debris falls from the valley walls and accumulates on top of the glacier. In addition, part of the load may be pushed along in front of the advancing ice.

Ice sheets are normally so thick that they gather very little debris to be carried as superglacial load. They are not usually capable of transporting as much material as a valley glacier, and most of the load is frozen into the bottom of the ice or shoved along ahead of it.

(6) Glacial Deposition

A glacier's load consists of rocks and soil randomly intermingled without regard to size, weight or composition. When the ice melts it will drop the debris, forming a variety of deposits which are designated as **glacial drift.** There are two types of drift: (*a*) **till,** glacial drift which has not been stratified or sorted by water action but has been deposited directly by the ice; and (*b*) **outwash** or stratified drift, composed of materials that have been sorted and deposited in definite layers by the action of water from the melting ice.

(*a*) **Till or Unstratified Deposits.** Laid down directly by the ice, deposits of till are composed of rock fragments of varying size, many of which are polished or bear glacial striae. **Boulder clay** is the most extensive of the drift deposits, consisting of large flat masses of thick tenacious clay with pebbles and boulders of all sizes. It outcrops on many of the cliffs and beaches around Flamborough (Yorkshire). Deposits of till form topographic features known as **moraines**—ridges or mounds of boulders, gravel, sand and clay deposited by a glacier. There are several types of moraines, each being named with respect to its relation to the glacier. A **terminal moraine** or **end moraine** is a mound of till formed at the terminus or end of a glacier. Terminal moraines mark the former position of the ice front. **Recessional moraines** are deposits of till left at various points as a glacier recedes or is temporarily stable. Irregular deposits of till deposited by retreating melting glaciers are called **ground moraines.** Drumlins are commonly developed on ground-moraine surfaces.

Each of the moraines described above is characteristic of both ice sheets and valley glaciers. The latter, however, produce two types of moraines that are not developed by ice sheets. Valley glaciers commonly display **lateral moraines,** which are ridges formed on each side of a glacial valley. They consist of material that has been eroded

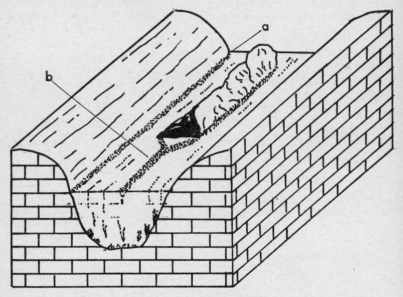

Fig. 81. Cross-section through valley glacier: (*a*) lateral moraine, (*b*) medial moraine

from the sides of the valley or has fallen from the valley sides on to the glacier's surface to be carried along by the ice. When two valley glaciers join to form a single stream of ice, the lateral moraines unite to form a single **medial moraine** (Fig. 81).

Erratics are large stones or boulders which have been transported by glaciers and are different from the underlying bedrock. Erratics, some of which may weigh many tons, have been found hundreds of miles from their place of origin. An elongated line of erratics derived from a common source is called a **boulder train.** Boulders of the distinctive Shap granite have been found strewn across the Pennines as far east as the Yorkshire beaches. At Robin Hood's Bay such erratics can be seen with erratics of Scandinavian origin, indicating that in this neighbourhood two glaciers once approached each other.

Sometimes the erratics are left perched precariously on top of the country rock.

(*b*) **Outwash or Stratified Deposits.** Rock materials deposited by streams of glacial meltwater are called outwash. Known also as **glaciofluvial** deposits, this stream-sorted material may produce a variety of land forms. Some of the more common outwash deposits are discussed below.

Outwash plains are broad, fan-shaped deposits of fine drift deposited in front of the glacier; they are characteristic of ice sheets. In a valley glacier most of the outwash deposits are left along the valley floor, forming features known as **valley trains.**

Eskers are long, winding ridges of stratified drift which resemble railway embankments. This material was apparently deposited in ice tunnels by streams which ran along the bottom of the glacier.

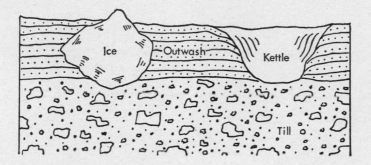

Fig. 82. Formation of kettle in outwash: block of ice (on the left) is buried in outwash; when ice melts, a depression (the kettle) remains

Although eskers may be many miles in length and from ten to as much as 100 feet high, they are only a few feet wide. The best examples of eskers in Britain are found to the north of Wolverhampton.

Kames, small, rather steep-sided, flat-topped hills of stratified drift, are formed by material that collected in circular depressions in the glacier. Layered deposits of sand and gravel formed between the side of a wasting glacier and the adjoining valley wall are called **kame terraces.** Kames are best seen in Scotland near Carstairs.

Kettles are depressions, some as much as a mile long and 100 feet deep, which mark the place where a block of ice (left by a retreating glacier) was buried in outwash. When the ice block melted the kettle was left to mark its place (Fig. 82).

Varve clays are clays which formed in the temporary lakes produced by ice in such features as kettles. Seasonal differences in material deposited are recorded in thin bands. They are occasionally found in Britain, but are best seen in Scandinavia, where they have been used extensively to unravel the mysteries of the Ice Age.

Crag-and-Tail. When a glacier passes over and round a resistant igneous plug or neck, the ice removes loose debris but deposits material behind it. The result is a hill having a steep face on one side and a gentle slope (the 'tail') on the other. A fine example of this can be seen at Edinburgh Castle. It is similar to, but much larger than, the roches moutonnées mentioned on page 79.

When the glacial ice melted in what are now temperate areas such as Britain, vast quantities of water were released and the river

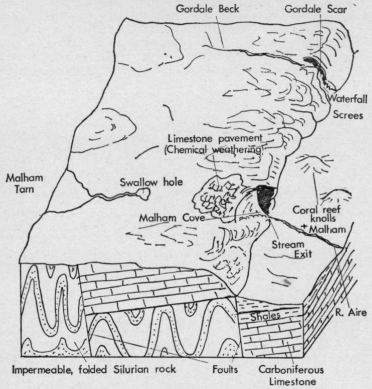

Fig. 83. Mallam Cove, Yorkshire

action discussed in Chapter Seven was accentuated. For example, huge boulders were hurled along through the underground passages in the limestone country. Underground waters met causing whirlpools, and great natural **chimneys** were carved out by the cork-screw action. Today some of the boulders are left incongruously and precariously perched high up in the roofs of limestone caves.

(7) Causes of Glacial Periods

What causes periods of glaciation? Although geologists have not yet found the complete answer to this rather complex problem, their studies of the record of glacial activities of the past, and observations on many of our present-day glaciers, have given rise to a number of theories. A few of the factors bearing on this problem are outlined below.

(*a*) **Elevation of the land.** Periods of glaciation seem to coincide with times when the continents are known to have been high. Lower temperatures prevail at higher altitudes, and a drop in mean annual temperature at a time when lands are high may produce a period of glaciation.

(*b*) **Variation in the amount of heat received from the sun.** The sun is the source of the earth's heat energy, and the amount of energy thus generated is known to have fluctuated as much as 3 per cent in the past 40 years. Although this is not enough to cause glaciation, larger fluctuations may have occurred in the geologic past. Variations in the amount of solar energy received by the earth's surface might also have been produced by occasional clouds of volcanic dust. Or perhaps the earth's orbit around the sun was different from what it is today, thus producing a longer, colder winter season.

(*c*) **Variations in carbon dioxide and water vapour in the atmosphere.** There is considerable evidence to suggest that both carbon dioxide and water vapour help the earth to retain heat derived from the sun. A decrease in these substances would permit more heat to escape by radiation, thus bringing about a colder climate. Enlarged and elevated land areas may have reduced the amount of water vapour in the atmosphere and thus decreased its ability to retain solar heat.

In addition to the above, such factors as volcanism, melting of the arctic ice cap due to changes in oceanic circulation, and shifting of the poles have been suggested as possible causes of glaciation.

(8) Work of the Wind

Wind (air in motion) is a very effective geologic agent. Although not normally as spectacular as the work of water or ice, wind erosion is, nonetheless, an important land-forming process. As might be expected, wind works most effectively in arid and semi-arid regions. The strange, $1\frac{1}{2}$-mile-long Ayers Rock in Australia has been furrowed, pocked and polished by wind action for over 200 million years. However, even moist regions experience periods of drought, and during such times the soil becomes loose and subject to removal by wind. In addition, wind-blown dust may be transported for great distances and deposited in areas in which climates are prevailingly humid. Dust storms originating in the Sahara have produced a fine deposit of sand as far north as Britain.

(9) Wind Erosion

Wind by itself has little, if any, effect on solid rock. But winds of high velocity will pick up a load of rock fragments which may become effective tools of erosion. These winds may erode by deflation or by abrasion.

Deflation. In regions of arid and semi-arid climates, and where the vegetative cover is sparse, loose rock and soil particles are apt to be blown away by the wind. Farming regions with fertile soils have been reduced to dust-bowls during droughts; basins, particularly in Egypt, some of which are well below sea level, have been created by deflation, exposing the bed rock and creating oases. Thus wind can have both a disastrous and a beneficial effect. This process of blowing away, known as deflation, may create several types of distinctive features. For example, broad shallow depressions termed **blow-outs** may be developed where wind has scooped out soft unconsolidated rocks and soil. **Lag gravels** are formed when the wind blows away finer rock particles, leaving behind a residue of coarse gravel and stones. **Desert pavement** may be produced when large pebbles, boulders, and patches of redrock are exposed in such a manner that

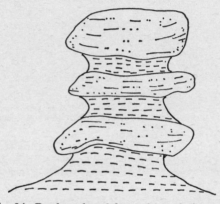

Fig. 84. Rock pedestal formed by wind erosion

they fit tightly together to form a relatively smooth surface. Some of these stones may exhibit a dark, enamel-like coat of iron or manganese called **desert varnish**.

Abrasion. The wind abrades by means of loose sand and dust particles which are transported as part of its load. Wind abrasion acts as a natural sandblasting process to wear away solid objects. The destructive action of these wind-blown abrasives may wear away wooden telegraph poles and fence posts, and scour or groove solid rock surfaces. Windows which are constantly exposed to the impact of wind-blown sand may eventually become pitted, chipped, or frosted. The sand grains which are used in the abrasive process are also subjected to wear; they, too, will become pitted, worn and reduced in size.

Wind abrasion also plays a part in the development of such land forms as **table rocks** and **pedestals**—isolated rocks which have had their bases undercut by wind-blown sand (Fig. 84). Moreover, certain shallow hillside caves appear to have been hollowed out with the help of wind erosion. **Ventifacts** are another interesting and relatively common product of wind erosion: they comprise pebbles, cobbles, and in some instances boulders which have been polished, faceted, or grooved by wind-blown sand. Ventifacts are formed when the wind blows sand against the side of the stone, shaping it into a flat surface (Fig. 85). If the prevailing winds change direction, faces may be developed on other sides of the stone. Stones formed with

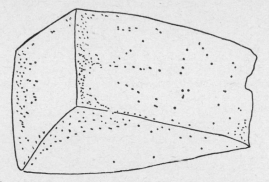

Fig. 85. Ventifact, an angular stone shaped by the wind

one flat face commonly have a single sharp edge and these are called **einkanters** (German, 'one-edge'); triangular-shaped, three-faced ventifacts are called **dreikanters** (German, 'three-edge').

Almost spherical grains of sand are produced in deserts. These **millet-seed** sands are devoid of mica, as it is powdered by wind action. This is a useful method of diagnosis of the origin of sandstones. Sandstones which are rich in feldspar and have no mica content were probably formed in arid desert conditions.

(10) Transportation by the Wind

The manner in which wind carries its load is determined by the size, shape, and weight of the rock particles and by the velocity of the wind. Wind-transported materials are most commonly derived from places containing loose, weathered rock fragments (for example, flood plains, beach sands, glacial deposits, and dried lake bottoms). In addition, volcanic explosions may produce large amounts of light ash or dust which will be transported by winds.

The wind is capable of transporting large quantities of material for very great distances. Part of the material rolls or slides along the ground. This, the **bed load,** is said to move by **traction.** Some sand particles move by **saltation**—a series of leaping or bounding movements. If the velocity of the wind is great enough, particles may be transported in **suspension.** Most of the suspended load is carried within a few feet of the ground, but lighter dust particles may be lifted upward into higher, faster-moving wind currents. Material caught up in these upper-level wind currents may be transported for many thousands of miles.

(11) **Wind Deposition**

The wind will begin to deposit its load when its velocity is decreased, or when the air is washed clean by falling rain or snow. A decrease in wind velocity may be brought about when the wind 'dies down' or strikes some obstacle (trees, fences) in its path.

The major types of wind-blown deposits (known also as **aeolian** deposits) are (*a*) **dunes** and (**b**) **loess.**

(*a*) **Dunes.** These are mounds or hills of sand which have been deposited by the wind. Dunes vary in shape and size according to the nature of the wind, available quantity of sand, and the amount and distribution of the vegetative cover.

Dunes are formed in areas where there is a sufficient amount of loose, unprotected sand and winds strong enough to move it. Areas of this type include sandy deserts, sandy flood plains, and sandy beaches along lake shores or sea coasts. The dune is started by an obstacle that causes a drop in the speed of the wind, perhaps a tree

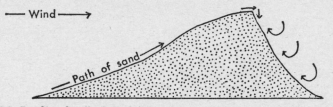

Fig. 86. Profile of typical sand dune; arrows denote path of wind currents

or a fence. When the velocity drops, a small deposit of wind-blown material will accumulate on the sheltered (leeward) side of the obstruction. As the mound of sand grows it becomes a more efficient windbreak and thus increases the amount of deposition. This process continues until the dune is at least several feet high, and might possibly continue until it is several hundreds of feet high.

Dunes formed in areas where the wind blows steadily from a single direction develop a characteristic profile (Fig. 86). Such dunes have a relatively long gentle slope on the windward side and a short steep slope on the leeward side. Small furrows, known as **ripple marks,** are commonly found on the windward slope of the dune.

Migration of Dunes. Most sand dunes are not stationary. Instead, they slowly migrate as the wind blows sand up the gentle windward slope and over the crest, thus allowing it to fall down the steep leeward side. The repeated occurrence of this process will result in the dune moving in a leeward direction, a movement called **dune migration.** Although normally a rather slow movement (seldom as much as 25 feet per year), some dunes are known to have moved as much as 100 feet in one year. Migration will continue until the dunes become covered with vegetation, which will protect the sand from the wind. Dunes of this type are said to be **fixed** or **stabilized.** Migrating dunes have been known to advance over forests, farm lands, railways, roads and villages. In some instances man has arrested the movement of sand by planting grass, shrubs, or trees or by erecting protective fences.

Types of Dunes. Dunes vary greatly in size and shape, depending on the velocity and direction of the wind and on the amount of sand available in the area.

Barchans. Crescent-shaped dunes characterized by two long, curved extensions pointing in the direction of the wind are called

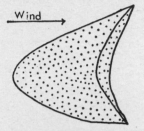

Fig. 87. Barchan sand dune

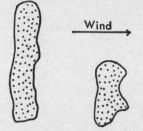

Fig. 88. Transverse sand dunes

barchans (Fig. 87). Such dunes are formed in areas where winds blow steadily and from a single direction.

Transverse Dunes. Formed especially along sea coasts and lake shores, transverse dunes develop with their long axis at right-angles to the wind (Fig. 88). Sand ridges of this type may be 10–15 feet high and as much as half a mile in length.

Longitudinal Dunes. A long ridgelike dune developed parallel to the wind is called a longitudinal dune (Fig. 89). **Seif dunes** are a

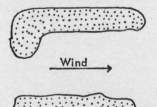

Fig. 89. Longitudinal sand dunes

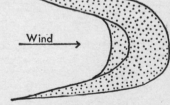

Fig. 90. Parabolic sand dune

special type of longitudinal dune which resemble the shape of an Arabian sword. Seif dunes may be as much as 700 feet high and 600 feet long. They are commonly grouped together to form ridges which may extend for many miles across the country.

Parabolic Dunes. These U-shaped dunes resemble barchans. However, the tips of parabolic dunes point toward the wind, whereas the tips of the barchan point downwind (Fig. 90).

(*b*) **Loess.** The finer particles carried by the wind may accumulate to form deposits of dust known as **loess.** A yellowish, fine-grained,

E

non-stratified material, loess is composed of small angular fragments of a variety of minerals. The materials forming loess are derived from surface dust originating primarily in deserts, river flood plains, glacial outwash deposits, and deltas. Loess is quite cohesive and possesses the property of forming steep bluffs with vertical faces.

Loess is well known for its ability to form fine-textured, fertile, yellowish soils, and in areas of sufficient rainfall such soils are of considerable agricultural importance. At the end of the Ice Age a belt of loess was deposited stretching from France to China. It starts as a thin deposit of glacial material in France, steadily thickening across Russia until it reaches its maximum thickness in China, where it is composed of desert products.

(12) Mass Movement of Rocks and Soil

Mass movement, or **mass wasting**, takes place as earth materials move down a slope in response to the force of gravity. This type of erosion is apt to occur in any area with slopes steep enough to allow downward movement of rock debris.

All land surfaces slope to a certain degree, and the ability of a slope to resist gravity depends largely on the cohesive ability or strength of the earth materials forming it. Some of the factors which help gravity overcome this resistance are discussed below.

Water. Although mass wasting may occur in either wet or dry materials, water greatly facilitates downhill movements. Water may soften clays, making them slippery; it adds weight to the rock mass; and, in large amounts, it may actually force rock particles apart, thus reducing soil cohesion.

Freezing and Thawing. Water contained in rock and soil expands when it freezes. Alternate periods of freezing and thawing will loosen rock materials, and in some instances ice expansion may be great enough to force rocks downhill. This action is most effective at high altitudes where freezing and thawing occur almost every day.

Undercutting. Undercutting by streams or by man-made excavations may remove support and allow overlying material to fall.

Organic Activities. Animals such as deer or cattle walking on the surface knock materials downhill. Burrowing animals cast rocks and soils out of their holes as they dig; these are commonly piled up at the foot of the slope.

Shock Waves. Strong vibrations caused by faulting, blasting, and heavy traffic can also exert sufficient stress on rock materials to start their movement downhill. An example of this occurred in Yellowstone National Park in August 1959, when a landslide was triggered by an earthquake.

Mass movements may occur suddenly and violently as in a landslide, or almost imperceptibly as in the case of soil creep. Let us now consider the various forms of rapid and slow mass movements.

(13) Rapid Movements

Most of the more rapid movements of earth materials are the results of forces which have been gradually weakening the mantle rock over long periods of time. Common examples of rapid mass wasting include the following:

Scree. Scree (or **talus**) consists of weathered rock fragments piled up at the foot of a cliff or mountain. Screes build up gradually as weathered rock particles are dislodged from the cliff face and roll downhill.

Landslides. The most spectacular and violent of all mass movements, landslides are characterized by the sudden downhill movement of great quantities of rock and soil. Such movements typically occur on steep slopes that have large accumulations of weathered material. Water from rain or snow may seep into the mass of steeply sloping rock debris, adding sufficient weight to start the entire mass sliding. **Avalanches** are rapid movements of snow with some soil and rocks. **Rock slides** are especially destructive landslides which involve movement of the bedrock as well as the mantle rock.

Slump. This special type of landslide occurs as large masses of a slope move downward and outward owing to gravitational pull. Such movements are most likely to occur in unconsolidated materials and are usually caused by undercutting or steepening of the slope to the extent that it can no longer support its own weight. Slump is a common occurrence along the banks of streams or the walls of steep valleys.

There are numerous accounts of destructive landslides which have taken place within living memory. One such catastrophe occurred in Alberta, Canada, in 1903 when some 40 million cubic yards of rock suddenly broke loose from the face of Turtle Mountain and descended on the little coal-mining town of Frank. This huge rock mass rushed two miles across the valley and 400 feet up the opposite side. The total period of movement was less than two minutes, but 70 residents of Frank lost their lives in that brief span of time.

Mudflows. Large flowing masses of rocks, soil and water mixed to a mudlike consistency are termed mudflows. Mass wasting of this type typically occurs as certain arid or semi-arid mountainous regions are subjected to unusually heavy rains. Originating as a rule in steep-walled gulches or canyons, mudflows course down the valley and can cause widespread destruction of objects in their path.

Slumps may be induced by the carelessness of man. The Aberfan tragedy in 1966 was a particularly disastrous case of a man-made slag-heap destroying a school and many of its pupils. Minor landslides are often induced when cuttings or embankments are constructed.

Earthflows. Differing from mudflows in the amount of water that they contain, earthflows usually move more slowly than the more fluid masses of mud.

(14) Slow Movements

Although the slower types of mass movement lack the sudden and spectacular action that marks rapid mass wasting, their total geologic effect is probably considerably greater.

Soil Creep. Gravity also moves rock material down a slope by means of soil creep. This movement, usually so slow as to be imperceptible, normally occurs on moist slopes not steep enough for landslides. As the mantle rock slowly moves downhill it may tilt trees,

Fig. 91. Soil creep. (Note downslope migration of strata, tilting the tree)

displace fences, and deform rock strata (Fig. 91). Soil creep may be accelerated by frost wedging (see page 77), by alternate freezing and thawing, and by certain plant and animal activities.

Solifluction. Solifluction is a downhill movement typical of areas where the ground is normally frozen to a considerable depth. The actual soil flow occurs when the upper portion of the mantle rock thaws and becomes water-saturated. The underlying, still frozen subsoil acts as a 'slide' for the sodden mantle rock which will move down even the most gentle slope. Solifluction is characteristic of arctic, sub-arctic, and high mountain regions.

CHAPTER NINE

OCEANS AND SHORELINES

The oceans of the world cover approximately 71 per cent of the earth's surface, and most of the soil which has been eroded from the land is eventually deposited in the sea. In addition to their geologic importance, oceans are of value to man as routes of commerce, regulators of climate, and the primary source of all water.

The oceanic waters of the earth have a combined area of about 150 million square miles (400 million km^2). In the southern hemisphere the oceans cover about 81 per cent of the surface, while in the northern hemisphere they cover approximately 61 per cent of the surface. The oceans are intercommunicating bodies of water; therefore a ship can sail from one ocean to all of the others. Oceans include also their adjoining gulfs and bays; thus, the Mediterranean and Baltic seas are considered to be parts of the Atlantic Ocean.

(1) Division of the Oceans

Geographers recognize five oceans: the Pacific, Atlantic, Indian, Arctic and Southern. The Pacific, the largest and deepest ocean, makes up approximately three-eighths of the total water area. At the equator, its widest part, the Pacific is about 10,000 miles (16,000 km) wide.

The second largest ocean, the Atlantic, comprises roughly one-quarter of the total area of the oceans and ranges from 2,000 to 4,200 miles (3,200 to 6,700 km) in width. The Indian Ocean is third in size, being about 6,000 miles (9,600 km) in diameter, and composing about one-eighth of the total sea area.

An extension of the Atlantic, the Arctic Ocean is from 1,500 to 3,000 miles (2,400 to 4,800 km) wide and constitutes about one-thirtieth of the total sea area. Its surface is covered by eight to ten feet of ice during most of the year. The remaining ocean waters comprise the Southern Ocean which surrounds the south polar Antarctic land mass.

(2) Depth of the Oceans

As noted above, the Pacific is the deepest of all oceans, having an average depth of about 14,000 feet (4,300 metres). At its deepest part, the Marianas Trench in the western Pacific, it is 35,800 feet deep. The Indian Ocean is second deepest, with a mean depth of approximately 13,000 feet, while the Atlantic Ocean has a mean depth of about 12,800 feet. The Arctic Ocean, which has an average depth of only 4,000 feet, is relatively shallow.

Those parts of the ocean floor which are more than 18,000 feet below sea level are called **deeps.** There are about 57 known deeps; Pacific deeps include the Marianas Trench off the island of Guam, and the Swire Deep or Philippine Trough (35,430 feet) located north-east of the island of Mindanao. The deepest part of the Atlantic Ocean, more than 29,000 feet (the height of Mount Everest is also 29,000 feet), is in the Milwaukee Deep, or Puerto Rico Trough, located just north of Puerto Rico.

The oceans are much deeper than the lands are high. The average depth of the sea is 13,000 feet (about $2\frac{1}{2}$ miles), while the average elevation of the land is about 2,600 feet (about half a mile). Thus, if the continents were eroded and the material composing them placed in the oceans, the earth would be covered by a universal sea approximately two miles deep.

(3) Composition of Ocean Water

Ocean water contains great quantities of dissolved gases (oxygen, nitrogen, and carbon dioxide) and mineral matter. Some of the more common solids normally to be found dissolved in sea water, their chemical symbols, and the relative amounts as percentages, are given below:

Sodium chloride	(NaCl)	77·76
Magnesium chloride	($MgCl_2$)	10·86
Magnesium sulphate	($MgSO_4$)	4·74
Calcium sulphate	($CaSO_4$)	3·60
Potassium sulphate	(K_2SO_4)	2·46
Calcium carbonate	($CaCO_3$)	0·34
Magnesium bromide	($MgBr_2$)	0·22

It is at once obvious that the primary solid constituents of ocean water are **salts**, and that sodium chloride (common salt) comprises more than three-quarters of the total dissolved solids. Where do these salts come from? The greater part of them have been leached from the soil and transported to sea by rivers.

The salt content of an ocean may vary from place to place. For example, waters may be saltier in regions where excessive evaporation occurs, and less saline in areas of cold climate or heavy rainfall, or where large rivers enter the ocean.

(4) Life in the Ocean

The sea abounds with countless plants and animals of many types. These organisms float, swim or crawl about, or burrow in the floor of the ocean. Others, such as certain calcium-secreting plants and corals, build great calcareous masses called **reefs.**

Most marine life exists in the shallow marginal or **epicontinental** seas that border the continents; many of our fossiliferous rock formations were deposited in similar seas in prehistoric time.

(5) The Ocean Floor

We know that parts of the ocean floor are marked by mountain ranges, plateaus, and other relief features similar to those of the land. In general, however, the submarine topography is not as rugged as that of the earth's surface.

The floor of the ocean has been divided into three parts: the continental shelf, continental slope, and deep-sea floor or abyssal zone (Fig. 92).

Continental Shelf. The flooded, nearly flat margins of the continents are called the continental shelves. Sloping gently outward from the shores of the continents, these shelves have an average width of about 40 miles and an average depth of about 400 feet.

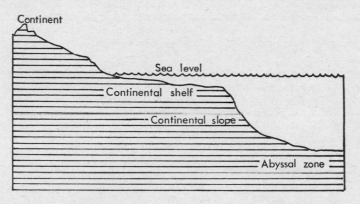

Fig. 92. Major divisions of ocean floor

Continental Slope. The slope of the sea floor increases rather sharply at the outer edges of the continental shelves. These slopes descend abruptly (some dropping as much as 30,000 feet in a relatively short distance) to the deeper parts of the ocean. In some places the surfaces of the continental slope and continental shelf are marked by deep submarine canyons. One of these, the Hudson Submarine Canyon, is about 2,400 feet deep, three miles wide, and about 125 miles long. The Hudson Submarine Canyon appears to be an extension of the valley of the Hudson River; other canyons, however, are not associated with extended river valleys and their origin is far from clear. Certain of these are believed to be the result of submarine earth movements, tidal scour (erosion produced by tidal action), changes in sea level during periods of glaciation, and turbidity currents (currents of turbid or muddy water moving relative to the surrounding water because of higher density).

Deep-sea Floor or Abyssal Zone. That part of the ocean floor extending seaward from the base of the continental slope is referred to as the deep-sea floor or abyssal zone. This floor is not flat; rather it is marked by mountain ranges, volcanic peaks, valleys, and deep basins. Among the more important features of the deep-sea floor are:

Abyssal plains. These are large flat areas having gradients of less than about five feet per mile.

Deep-sea trenches. Also called ocean **deeps,** these are long, narrow, basins in the deep-sea floor. Many occur at the foot of the continental slope, and although their origin is not fully known, they may be associated with submarine faulting.

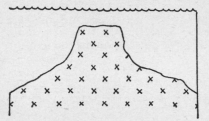

Seamounts. Isolated mountain-shaped elevations more than 3,000 feet high are called seamounts. These may occur on the mid-ocean ridges (narrow, steep-sided elevations on the ocean floor) as well as on the deep-sea floor.

Fig. 93. A guyot; such structures are typically covered by 3,000 to 6,000 feet of water

Guyots. These are flat-topped seamounts (Fig. 93) rising from the ocean bottom and usually covered by 3,000 to 6,000 feet of water. Especially well known in the Pacific, guyots could be submerged volcanoes which have been truncated by wave action.

(6) Movements of the Sea

Anyone who has witnessed the ceaseless, restless motion of the sea can easily understand its effectiveness as a geologic agent. Tides, currents, and waves, the principal types of movements of ocean water, are continually at work producing changes in the rocks along the shore.

The causes of the relentless motion of the sea are varied and complex. Basically, however, they may be attributed to tides, wind, changing density of sea water, and rotation of the earth.

Tides. The periodic rise and fall of the sea (once every 12 hours and 26 minutes) produces the tides. Tides cause the ocean waters to rise gradually for about 6 hours and 13 minutes and to recede slowly for an equal period of time. The effect of the tides is only slight in the open sea, the difference between high and low tide (the **tidal range**) amounting to about two feet.

The tidal range may be considerably greater near shore. It may range, for example, from less than two feet in the Mediterranean to 30 feet in narrow stretches of water, as in the Mersey district and the Bristol Channel where spring tides may exceed 40 feet in height, while in other parts of the world tides of over 50 feet have been reported. The exceptionally high tides in the Bristol Channel are produced by the **tidal bore,** an advancing wall of water created when tides move into narrow bays and river mouths. The tidal range will vary according to the phase of the moon and the distance of the moon from the earth. The type of shoreline and the physical con-figuration of the ocean floor will also affect the tidal range.

Most tides are caused by the gravitational pull of the moon on the waters of the earth; the gravitational force exerted by the sun may also raise tides. In addition, the tidal effects of both the sun and the moon may be aided by centrifugal force developed by the rotation of the earth.

Currents. The oceans have localized movements of masses of sea water called ocean currents. These may be caused by prevailing winds, tides, variations in salinity of the water, rotation of the earth and concentrations of turbid or muddy water. Changes in water density due to temperature changes will also cause currents. The Gulf Stream, known as the North Atlantic Drift where it approaches the north-western shores of Europe, is a current of this type. The Gulf Stream is responsible for Britain's relatively mild winters and brings with it to these shores, and encourages, a rich fauna.

In addition to the ocean currents, there are several types of shore currents which are restricted to coastal regions. These include: **undertow,** a stream of water which returns seaward along the bottom underneath the incoming waves; **rip currents** (also caused by water returning to the ocean), swift currents running close to and parallel to the shoreline; and **longshore currents,** caused by waves which strike the shore at an oblique angle, creating currents parallel to the shoreline. Longshore currents are of considerable importance in shaping shorelines (see page 126).

Waves. Waves are produced by the friction of wind on open water. Essentially an up-and-down movement of the water, wave motion also moves the surface water in the direction that the wind is blow-ing. Breakers are formed when the wave comes into shallow water near the shore. The lower part of the wave is retarded by the ocean bottom, while the top, having greater momentum, is hurled forward causing the wave to break.

Varying greatly in size, waves may be slight ripples or giant storm waves 25–50 feet in height. The latter may cause severe damage to property as they race across coastal lowlands driven by gale-force winds or hurricanes. Giant waves may also be produced as a result of earthquakes on the ocean floor. These waves, called *tsunamis,* are the largest and most destructive of all ocean waves. They are discussed in Chapter Eleven.

The geologic work of the sea, like that of water, ice and wind, consists of erosion, transportation and deposition. The sea accom-plishes its work largely by means of waves and wave-produced currents; their effect on the coastline may be quite pronounced.

(7) Marine Erosion

As waves attack the shore they erode by a combination of several processes. How rapidly the shore is worn away depends, of course, on the resistance of the rocks composing it and on the intensity of wave action to which it is subjected. In exceptional cases house owners who had bought houses with pleasant gardens near the sea have found their gardens diminishing yearly and their homes gradually breaking up and sliding into the sea.

Processes of Marine Erosion. Wave erosion may be accomplished in a variety of ways. Destruction by *hydraulic action* occurs as waves beat on poorly consolidated sediments or loosely jointed rock. The coast also undergoes *abrasion* as rock fragments carried by waves and currents scour and grind against it. Coasts having outcrops of soluble rocks such as limestone may also be affected by *solution*.

Features Formed by Marine Erosion. Some of the most common features of coastal scenery are the result of wave erosion. For example:

Sea Cliffs. Known also as **wave-cut cliffs,** sea cliffs (Fig. 94) are formed by the caving-in of overhanging rocks after the underlying

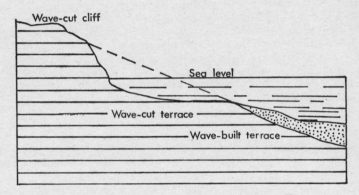

Fig. 94. Features formed by marine erosion and deposition

rock has been eroded by waves. Such cliffs are essentially vertical and are quite common. The best known in Britain are the white chalk cliffs of Dover. The Saw-Tooth Cliffs around the Great Australian Bight, 720 miles in length, are the world's longest unbroken stretch of cliffs.

Wave-cut Bench. Extending seaward from the base of most sea cliffs are relatively flat platforms called wave-cut benches or wave-cut terraces (Fig. 94). Such benches are a feature of the Yorkshire coast between Whitby and Robin Hood's Bay. Also found on these beaches are features similar to the wind-eroded pedestals mentioned in Chapter Eight. In this case the dark shales have been worn away leaving caps of more resistant material standing up like mushrooms.

Headlands or Promontories. Finger-like projections of resistant rock extending out into the water are known as headlands or promontories (Fig. 95). The relatively resistant chalk of Culver Down on the Isle of Wight stands out as high ground forming a headland between Whitecliff Bay and Sandown Bay. In the former, soft

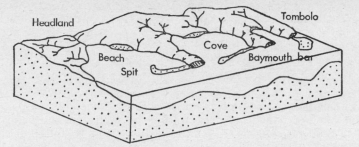

Fig. 95. Shoreline features caused by marine erosion and deposition

Eocene rocks have been rapidly eroded, and in the latter, the same has happened to the older Cretaceous beds. Indentations between headlands are termed **coves**, one of the examples of which is Lulworth Cove in Dorset.

Sea Caves, Sea Arches and Stacks. Continued wave action on a sea cliff may hollow out cavities which will form **sea caves.** Fine examples can be seen at Flamborough Head in Yorkshire, where also the beach shows a particular form of erosion known as **solution hollows.** The wave-cut beach is pocked by chemical solution of the chalk forming craggy ruts and holes. Waves may cut completely through a headland to form a **sea arch**; should the roof of the arch collapse, the rock left separated from the headland is called a **stack,** such as The Needles off the Isle of Wight and Old Harry in Dorset. These features are shown diagramatically in Fig. 96.

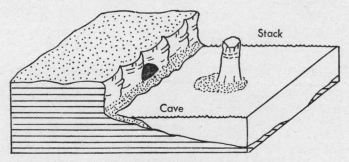

Fig. 96. Coastal features developed by marine erosion

(8) Marine Transportation

Waves and currents are not only the major agents of marine erosion, they are also important transporting agents. Undertow and rip currents carry rock particles back to the sea; longshore currents pick up sediments, moving them out from shore into deeper water. Some of this material is carried in solution. Materials carried in suspension or solution may drift seaward for great distances and eventually be deposited far from shore. While being carried by waves and currents, sediments undergo additional erosion, becoming rounded and reduced in size.

(9) Marine Deposition

When waves or currents suffer reduced velocity they will deposit their load. In addition, some rock particles may be thrown up on the shore by the breaking waves. Most of the sediments thus deposited consist of rock fragments derived from the mechanical weathering of the continents, and they differ considerably from terrestrial or continental deposits.

Features Formed by Marine Deposition. While erosion may be taking place along one part of the coast, marine sediments are being deposited elsewhere. Deposits formed between high tide and low tide levels are known as **littoral deposits.** These depositional features, like those formed by marine erosion, are characteristic of most shorelines. For example:

Beaches. Beaches are coastal deposits of debris which lie above the low-tide limit in the shore zone. They are transitory features; and although most beaches are sandy, they may also consist of pebbles, cobbles, shells, mud, or a combination of these materials.

Offshore (or barrier) bars. Long narrow accumulations of sand lying parallel to the shore and separated from the shore by a shallow lagoon are called offshore bars or barrier bars. The famous Chesil Beach off the Dorset coast is a shingle bar extending for about 16 miles from Bridport to Portland. It is composed of pebbles from local rocks and from as far afield as Devon and Cornwall.

Spits and Hooks. Long, narrow embankments of sand and pebbles extending out into the water but attached by one end to the land are called spits (Fig. 95). When the free end of a spit curves landward a 'hook' or 'recurved spit' is formed, such as Calshot Spit and Hurst Spit in the Solent.

Tombolos. A tombolo is a bar of sand or gravel connecting an island to the mainland or to another island (Fig. 95). Islands associated with tombolos are known as **tied islands.** The Isle of Portland is a tied island connected to the mainland near Weymouth by a tombolo. Gibraltar is also a tied island. Alexander the Great in 330 B.C. built a 650-yard artificial tombolo (causeway) from the rubble of a crushed city out to the island site of ancient Tyre, but this has long since been silted over. (Some geologists consider a tombolo to consist of both the island and the connecting bar rather than just the bar as defined above.)

Wave-built Terraces. Sediments deposited in deep water beyond a wave-cut terrace may form deposits referred to as wave-built terraces.

A classic example of this is in south-west Ireland from Cork to Kerry. Other examples are the ria coasts of north-west Spain and the fiord coasts of Norway and New Zealand. The prominence of this type of coast is a direct result of the changes experienced in the post-glacial period.

(10) Shoreline Development

Shorelines are developed over long periods of time as the result of changes in sea level aided by wave and current erosion. Although several classifications of shorelines have been proposed, geologists still cannot reach agreement as to which of these should be adopted. Two of the more commonly used classifications are briefly outlined below.

(*a*) **Johnson's Classification.** First proposed in 1919 by Prof. D. W. Johnson of Columbia University, this scheme is based on the relative movement of the land with respect to sea level; in other words, has the coast been elevated, or has it subsided? The four types of shorelines recognized in this classification are:

(i) *Shorelines of Submergence.* Submergent shorelines result from a sinking land mass or a rising sea. Coasts of this type are typically deep, irregular in outline, and characterized by many headlands, coves, and drowned valleys which have become bays or estuaries. Islands, sea cliffs, stacks, bars, and tombolos are common along such coasts.

(ii) *Shorelines of Emergence.* Shorelines which have recently emerged exhibit a regular outline, offshore bars and lagoons, and relatively few bays. Flat coastal plains, representing a raised portion of the old sea floor, are also characteristic of emergent shorelines. Such coasts are not common as uplift has to overcome the effect of submergence; hence the most abundant forms of shoreline are compound shorelines, described below. Finland's shoreline is approaching the state of an emergent shoreline, but the isostatic readjustment (see page 75) has not yet been completed. The best examples of such a shoreline occur along the coast of Texas in the Gulf of Mexico.

(iii) *Neutral Shorelines.* Shorelines that are neither submergent nor emergent are termed 'neutral' shorelines. They may be formed by deltas constructed at river mouths, by outwash plains in glaciated areas, by lava flows in volcanic areas, and by coral reefs.

(iv) *Compound Shorelines.* Some shorelines display the characteristics of both submergence and emergence. These compound shorelines have rather complex geologic histories and have normally undergone periods of both submergence and emergence. Much of the Atlantic coast of the United States, from Virginia to Florida, is bordered by a shoreline of this type.

(*b*) **Shepard's Classification.** A later classification, proposed in 1937 by Prof. F. P. Shepard of the Scripps Institution of Oceanography (U.S.A.), overcomes some of the problems of Johnson's classification. Although this classification is quite comprehensive and has drawn the support of many oceanographers and geologists, it is still difficult to apply to certain types of coasts. In essence, the Shepard classification is as follows:

(i) *Primary*, or *youthful*, shorelines, shaped mainly by non-marine agencies:
   (*A*) Coasts shaped by terrestrial erosion and drowned by down-warping or deglaciation
      1. Ria coasts (drowned river-valley coasts)
      2. Drowned glaciated coasts
   (*B*) Coasts shaped by land-deposited materials
      1. River-deposition coasts
         (*a*) Delta coasts
         (*b*) Drowned alluvial-plain coasts
      2. Glacial-deposition coasts
         (*a*) Partially submerged moraine
         (*b*) Partially submerged drumlins
      3. Wind-deposition coasts
      4. Vegetation-extended coasts
   (*C*) Coasts shaped by volcanic activity
      1. Volcanic deposition (recent lava-flow coasts)
      2. Volcanic explosion or collapse
   (*D*) Coasts shaped by diastrophism (movements within the earth's crust)
      1. Fault-scarp coasts
      2. Coasts due to folding

(ii) *Secondary*, or *mature*, shorelines, shaped primarily by marine agencies:
   (*A*) Coasts shaped by marine erosion
      1. Sea cliffs straightened by marine erosion
      2. Sea cliffs made irregular by marine erosion
   (*B*) Coasts shaped by marine deposition
      1. Coasts straightened by deposition of bars and spits
      2. Coasts prograded (built outward) by marine deposition
      3. Coasts having offshore bars and longshore spits
      4. Coral-reef coasts

The above classification was designed to apply to smaller subdivisions of ocean shorelines. Shepard proposed the following broad subdivisions to classify the larger coastal regions of the world:

   (i) Coasts with young mountains (mountains formed during Tertiary or Quaternary time—see Chapter Eighteen)
   (ii) Coasts with old mountains (mountains formed prior to Tertiary time)
   (iii) Coasts with broad coastal plains
   (iv) Glaciated coasts

Like many other natural phenomena, shorelines are so varied and complex that no one classification is wholly satisfactory. Moreover, shoreline classification is further complicated by the many changes that a coast may undergo during its development.

(11) Coral Reefs

Coral reefs, previously mentioned as a factor in the formation of neutral shorelines or mature shorelines shaped by marine deposition, are ridges of calcareous rock at or near the surface of the sea. They consist, in part, of great accumulations of the limy skeletons of reef-building corals. These small animals, found only in the sea, generally live in colonies in warm (no colder than 68°F), clear water at depths of less than 150 feet. Corals of this type extract dissolved

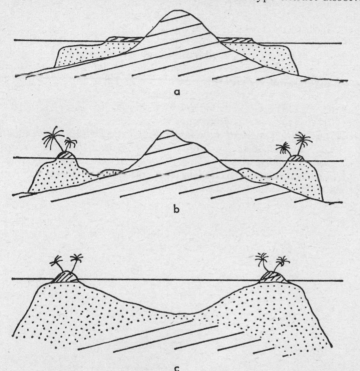

Fig. 97. Sequence of coral-reef formation according to subsidence theory of Darwin: (*a*) fringing reef, (*b*) barrier reef, (*c*) atoll

calcium carbonate from the sea water and use this material to build their shells. Other calcium-secreting plants and animals also live in the reefs. When these organisms die, their remains are added to the calcareous mass and new corals grow on top of them. Thus, the reef continues to grow for long periods of time.

The largest known coral reef, the Great Barrier Reef off the east coast of Queensland, Australia, extends for a distance of more than 1,200 miles and varies from ten to 90 miles in width. It is separated from the mainland by a wide lagoon, the Inland Water Way, which is an important route of commerce. Fossil coral reefs have been found in many parts of the world in rocks of various ages. Examples of fossilized reefs in Britain occur mainly in rocks of Silurian to Carboniferous age and in the Jurassic, indicating a warmer climate in times past.

Geologists and oceanographers have long speculated as to how the various types of coral reefs and coral islands have been formed. One of the more commonly accepted theories was proposed by Charles Darwin in 1842. According to his **Subsidence Hypothesis,** there are three stages of coral-reef formation (see Fig. 97). At first, a fringing reef is formed as corals grow in shallow water near the shore of an island. With the passage of time the island gradually subsides, while the corals continue to grow on top of the reef. Gradually the reef enlarges; it becomes separated from the shrinking island by a lagoon, and a barrier reef is created. The final stage, in Darwin's hypothesis, results in an **atoll,** a roughly circular reef surrounding a lagoon which covers the now submerged island.

A later idea, the **Glacial-control Hypothesis,** proposed in 1910 by Prof. R. A. Daly of Harvard, suggests that barrier reefs and atolls were formed on truncated volcanic islands as a result of changes in sea level during the Ice Age. Unfortunately, neither of these hypotheses satisfactorily explains all coral-reef structures. It has been suggested, therefore, that coral reefs may be formed as a result of both subsidence and glacial control.

CHAPTER TEN

LAKES AND SWAMPS

A lake may be defined as a body of standing water occupying a depression in the land. Varying greatly in size, lakes may range from less than an acre to many thousands of square miles in area. They also vary greatly in depth; some lakes are only a few feet deep and may dry up during periods of drought, while others are many thousands of feet deep. The world's largest lake is the Caspian Sea, with an area of about 169,000 square miles. (The term 'sea' is misused in this sense, for such bodies of water as the Dead Sea and Sea of Galilee are actually saline lakes.) The world's deepest lake, Lake Baykal in eastern Siberia, has a depth of more than 5,400 feet.

Lakes are to be found at all altitudes. Lake Titicaca, located between Chile and Peru, is 12,500 feet above sea level. At the other extreme, the Dead Sea in Israel and Jordan is almost 1,300 feet *below* sea level.

Some lakes, especially the larger ones, have considerable influence on the population around them. They may, for example, supply water for drinking and industrial purposes, regulate the temperature of the adjoining lands, provide recreational areas attracting large numbers of tourists (especially when they are associated with mountain scenery as in the Lake District), and be used as transportation routes. Because of this it is not uncommon to find that large cities and industrial areas have grown up around many of the world's larger lakes.

(1) Origin of Lake Basins

Lake basins may be formed in a variety of ways and by a number of different geologic processes. The more important of these are discussed below.

Crustal Movements. Some basins are the results of warping, folding or faulting of the strata. For example, Lake Superior occupies a basin that was originally formed by structural deformation and later enlarged by glacial action. **Rift-valley lakes** are created when great fault blocks (see Chapter Five) sink down between high steep walls. Usually quite deep, lakes of this type often form a chain along the floor of the valley. Such a chain exists in the Great Rift Valley that extends from the Dead Sea to Lakes Nyasa and Tanganyika in eastern Africa.

Lakes are sometimes formed as a result of rock displacement which accompanies earthquakes; a local sinking of the ground may produce depressions suitable for lake formation. Reelfoot Lake in north-western Tennessee was formed in this way after the New Madrid earthquake of 1811.

Volcanic Activity. The effects of volcanism may form lakes of two kinds. Lava flows may dam up part of a stream valley, causing the stream to back up and form a lake. The Sea of Galilee was formed in this way. The other kind of lake is that formed in the craters or calderas of extinct volcanoes. The best-known example of this type is Crater Lake in the Cascade Mountains of south-western Oregon. The lake, which occupies a caldera, is about six miles wide and 2,000 feet deep, and is surrounded by cliffs 500 to 2,000 feet high.

Glaciation. Large numbers of lakes have been formed by glacial action. Such lakes may occupy surface depressions scooped out by glacial erosion, or basins developed behind natural dams of ice-deposited material.

Lakes formed at the head of glacial valleys are called **tarns** (see Chapter Eight) and these are fairly common in North Wales and other mountainous regions. Even more common are lakes occupying depressions scoured out by glaciers. Many of the elongated lakes of North Wales and the Lake District, and some of the corrie-lochs of Scotland, owe their origin to glacial scour. Other lochs have formed in the fault-block belts of the Grampians. Glacial lakes are found in many parts of the world: Dove Lake in Tasmania and the lakes high up in the Snowy Mountains of Australia are but two examples.

The Vale of York was once one large glacial lake; now, all that remains of Lake Humber are swamps on the moors. Lake Pickering, also formed by 'ponding up' during the Ice Age, overflowed and scooped out a gorge into which the Derwent still flows, strangely turning inland rather than out into the North Sea. Its remnants exist as marshes and peat bogs.

Mass Movements or Gravity. Lakes are sometimes created when earth debris from a landslide plunges down into a river valley, blocking the course of the river. The natural dam formed in this manner may cause the water to back up, creating a natural lake. A large lake once existed on the Upper Rhine near Flims due to such a landslide. A present-day example is Lake San Cristobal in Colorado which formed behind a large mudflow.

Rivers. Lake basins can also result from the erosional and depositional work of rivers. They may develop when cut-off meanders form ox-bow lakes (see Chapter Seven), such as those along the flood plain of the Mississippi River. They may also develop on deltas—Lake Pontchartrain on the Mississippi delta is a good example.

Ground Water. Ground water may form sink-holes and caverns in areas underlain by soluble rocks such as limestone and dolomite. If the sinkholes become choked with debris they may become filled with water, thus forming a lake. There are numerous lakes of this type in Kentucky and elsewhere. Small lakes are found in Cheshire where the ground has subsided owing to the solution of rock-salt.

Waves and Currents. Lakes are not uncommon along coastal plains or near certain lake and ocean shorelines. These are generally formed when waves and currents deposit sand bars across the mouth of a bay or lagoon. Originally salty or brackish, the water of these shoreline lakes may eventually be replaced by fresh water.

Other Causes. Man has created a number of artificial lakes and has increased the size of natural ones by building dams across the path of rivers. Derwent Water in Cumberland and the reservoirs of the Peak District of Derbyshire are beautiful examples of this. In addition, beavers' dams across streams, swamp vegetation, and log jams have been known to form lakes.

A few lakes have developed in craters created by meteorites. Water-filled craters of this type are known in northern Quebec and northern Siberia. Lakes also form in disused quarries, behind resistant dykes, and through the action of wind (as in the Landes in France).

(2) Types of Lakes

Lakes are generally classified as either **fresh-water** or **saline,** depending on the composition of their waters.

Fresh-water Lakes. Lakes that have outlets are called fresh-water lakes. The outlet is normally a surface stream, but some water may escape by seepage. The water in a fresh-water lake is derived from rain, melting snow, ground water, and river water. The composition of the water in the lake will depend largely upon the composition of the rocks over or through which the water has passed.

Although some of the dissolved material brought in by streams may be deposited in the lake, most of it leaves by way of the outlet. This, plus the addition of rain and snow, tends to keep the lake water fresh.

Most of the earth's lakes are of the fresh-water type. The largest of these, Lake Superior, has an area of 31,820 square miles.

Saline or Salt Lakes. Lakes that have no outlet will become saline or salt lakes. These are normally developed in dry climates where lakes lose their water by evaporation. As the water evaporates, the mineral salts become highly concentrated and may eventually be deposited on the lake bottom by precipitation. Because the mineral composition of the water is determined by the rock with which it has been in contact, the salts present will differ according to the type of rock in and around the lake basin.

Some salt lakes, such as the Caspian Sea, have been formed when portions of the sea became blocked off and isolated. In these cases, the waters were, of course, originally salty. Others, for example Great Salt Lake in Utah, started as fresh-water bodies, as did the Dead Sea. Such lakes may be seven or more times saltier than the world's oceans.

Salt lakes whose waters are characterized by large amounts of sodium carbonate or potassium carbonate are called **alkaline lakes.** The alkaline materials are commonly derived from high-sodium-content igneous rocks such as granite. Mono Lake in California is an example of an alkaline lake.

Playa Lakes. Usually developed in the lowest part of a desert

basin, playa lakes are shallow temporary lakes formed after heavy
rains. Playas disappear during dry periods, leaving behind a deposit
of silt or clay that may be covered with salts. These salt-covered dry
lake beds are called **alkali flats** or **salinas.** Typical of arid and semi-
arid regions in many parts of the world, good examples of playas
are found in Australia.

(3) Destruction of Lakes

Lakes are relatively temporary features, and the same geologic
processes that create lakes may also bring about their destruction.
Some lakes are destroyed because their basins become filled with
sediments or because the adjacent high land (or **rim**) is removed by
erosion. This process may be seen in action at the head of Crummock
Water in the Lake District. Lakes will also disappear when they lose
their water through excessive evaporation, seepage from below, or
diversion of the streams that flow into them.

Some lakes become filled with organic material. Vegetation such
as mosses, ferns and swamp grass which grow along the borders of
the lake may gradually spread towards the centre. The remains of
animals may also be mixed with the plant material, and this organic
debris may ultimately turn the lake into a **swamp.**

In addition, landslides, clay deposited by melting glacial ice, and
wind-blown sand and volcanic ash may also contribute to the
destruction of lakes. Lake Humber and Lake Pickering are examples
of extinct lakes.

(4) Swamps

Depressions that are partially or completely filled with living and
decomposed plant materials, sediments and water are called swamps.
Many swamps have been formed from the lake-filling processes
described above; others are simply areas of low, soft, boggy ground
which do not have proper surface drainage.

Known also as **bogs**, swamps commonly occur on the flood plains
of old rivers, such as the Mississippi. Romney Marsh, an **estuarine
marsh,** was formed by the silting up of a large area of south-east
Kent. They are also found on coastal plains, where they are commonly referred to as **marshes** or **tidal marshes.** Many of these occur
on the Gulf and Atlantic coasts of the United States.

Certain swamps are characterized by great thicknesses of partially
carbonized plant material called **peat,** notably found in Ireland. The
material dug from these peat bogs has a high carbon content and
when dry can be used as fuel. The process by which the swamp
vegetation is converted into peat is called **carbonization** (see Chapter
Fifteen), and the development of peat is the first step in the formation of coal.

Swamps also occur in glaciated regions. Here streams have been
blocked by glacial deposits which have created both lakes and
swamps (as in the Vale of York).

In areas where the ground is permanently frozen at shallow depths, swamps may develop on the upper surface during times of thaw. This spongy, waterlogged upper surface is known as the **tundra** and is well developed in the arctic regions of North America, Europe, and Asia.

CHAPTER ELEVEN

EARTHQUAKES AND THE
INTERIOR OF THE EARTH

Earthquakes, natural vibrations within the earth's crust, provide indisputable evidence that crustal movements are still taking place today. Some of these earth tremors are quite violent and are responsible for large-scale death and destruction. Most, however, are too small to be felt by man and must be detected by means of delicate recording instruments called **seismographs** (see Fig. 99).

(1) Causes of Earthquakes

Man has speculated about the origin of earthquakes for centuries; the ancients explained them as evidence of God's displeasure with the world, or the restlessness of some animal upon whose back the earth was resting.

Although **seismology** (the study of earthquakes) has provided us with much information about earthquakes, their ultimate cause is still not known with certainty.

We do know that earth tremblings are initiated by a sudden jar or shock and that most such shocks appear to be associated with faulting. The sudden fracture and displacement of the rocks along the fault plane generates a wavelike motion in the rocks. The manner in which these fractured rocks react is best explained by the **elastic rebound theory**. According to this theory, sub-surface rock masses subjected to prolonged pressures from different directions will slowly bend and change shape. Continued pressures set up strains so great that the rocks will eventually break and suddenly snap back into their original unstrained state. It is the elastic rebound (snapping back) that generates the seismic (earthquake) waves.

Earthquakes caused in the manner described above are called **tectonic** earthquakes and are the largest and most destructive of all. Earthquake waves may also be generated by volcanic activity. These may be caused by violent explosions (see Chapter Three) or by sudden movements of molten rock below the surface. Minor causes of earth tremors include rapid mass movements such as landslides or avalanches, and the sudden collapse of caverns.

(2) Distribution of Earthquakes

Although earthquakes may occur at any place over the entire earth, most of them originate in areas of crustal unrest and are associated with mountain-building movements. Earthquakes, like

136

volcanoes, occur in rather well-defined seismic belts (Fig. 98). About 80 per cent of the world's earthquakes originate in the Circum-Pacific belt of young mountain ranges and chains of volcanic mountains. This belt extends from Chile along the western borders of South and North America, northward to the Aleutians, Alaska, Japan, the Philippines, Indonesia, New Zealand, and certain Pacific islands. The second major seismic belt, the Mediterranean and Trans-Asiatic belt, extends from the Caribbean area through the Himalayas and Alps and includes Spain, Italy, Greece, and northern India. Approximately 15 per cent of the earth's seismic energy is released in the Mediterranean and Trans-Asiatic zone; the remaining 5 per cent is released in other parts of the world.

Britain does not lie in either of the two major seismic belts and therefore rarely suffers from earthquakes. Scotland occasionally

Fig. 98. Shaded areas designate the world's major seismic belts

experiences minor tremors as slight movements take place along old fault lines, such as those which separate the Highlands from the Midland Valley. A minor earthquake took place in the south of England in the early 1960s, disturbing the sleep of people in Hampshire. Sometimes earthquakes in other parts of the world produce a noticeable effect in Britain. The Lisbon earthquake of 1755 caused exceptional tides and produced rapid rises and falls of over two feet in some of the Scottish lochs. As we shall see, Britain is extremely fortunate not to be in an earthquake zone, but this has not always been so; the pattern continuously, although almost imperceptibly, changes. In past geologic times the British Isles must have suffered from earthquakes every bit as severe as those experienced in some parts of the world today. The feeble movements in Scotland are but the tail-end of a series which must have shaken the area in times past.

(3) Effects of Earthquakes

The destructive effects of earthquakes are familiar to almost everyone: shattered buildings, displaced roads and railways, collapsed bridges, great cracks in the ground and changes in sea level.

These are but a few of the physical changes that may be brought about by earthquakes. In some earthquake-stricken cities there has been more damage from fire than from seismic waves. Disrupted gas mains and electrical cables and overturned stoves may start serious fires. Attempts to extinguish such fires are usually handicapped by the loss of fire-fighting equipment, and broken water mains and communication lines.

The loss of life accompanying a major earthquake may be staggering. For example, an earthquake that occurred in north-central China in 1556 is said to have killed 830,000 people. Because earthquakes normally occur rapidly and unexpectedly, there is little time for precautionary measures. The death toll may also be raised by complicating factors such as disease, flood, fire and famine.

In addition to the loss of life and property, an earthquake will usually produce numerous geologic changes in an area. These include landslides, avalanches and mudflows, disruption of ground-water circulation, and sunken and fissured ground.

Earthquakes that occur beneath the ocean often generate great waves of water called **tsunamis** or **seismic sea waves.** These waves, which have been known to be as much as 200 feet high and to travel at speeds of up to 500 miles per hour, are capable of producing tremendous destruction. One such wave, associated with the Lisbon earthquake of 1755, attained an estimated height of 50 feet and washed inland for more than half a mile. Another, occurring north of Tokyo along the Pacific coast of Japan in 1896, reached heights of up to 100 feet and was responsible for the deaths of more than 27,000 persons.

(4) Historic Earthquakes

History contains numerous accounts of earthquakes, some of which go back before the birth of Christ. A few of the more serious instances are briefly described below.

Lisbon, Portugal (1755). This was perhaps the strongest earthquake in recorded history. Although the initial seismic shock lasted for only six or seven minutes, it destroyed about half of the city and was felt over an area of 1,250,000 square miles. The first tremor was followed by two other severe shocks; one 20 minutes after the first, and another some two hours later. The giant seismic sea wave accompanying this disaster advanced inland for more than half a mile, destroying almost everything in its path. Widespread fires added to the great property damage (the cost of which ran into millions of pounds) and the death toll has been estimated as high as 60,000 people.

New Madrid, Missouri (1811–12). This series of seismic shocks was felt from the Atlantic coast of America to the Rocky Mountains, and from Canada to the Gulf of Mexico. The first shock, which occurred at about 2 a.m. on December 16th, 1811, was followed by a series of after-shocks which continued for days. Another very severe shock occurred in late January 1812; the third shock, which was the most severe of all, took place in early February. During this three-month period of seismic disturbance as many as 1,874 shocks were recorded in Louisville, Kentucky, 200 miles away. Fortunately, this part of the United States was sparsely settled during the early nineteenth century, so there was little loss of life or property. The region did, however, undergo considerable geologic change. Landslides drastically affected the topography, and even changed land boundaries. Areas that were originally swamps were uplifted and drained; other areas sank from three to ten feet to form new swamps and lakes, and the course of the Mississippi River was changed. Reelfoot Lake in north-western Tennessee was one of the larger lakes thus formed; it is about 18 miles long and several miles wide.

Charleston, South Carolina (1886). This disturbance, like the New Madrid shocks described above, took place in an area which is generally considered to be stable. During the earthquake, the seismic waves increased in intensity until the ground seemed to rise and fall in visible waves. This earthquake, which did considerable damage and killed 27 persons, was felt over most of the eastern part of the United States.

San Francisco, California (1906). This well-known and severe earthquake occurred in the early morning of April 18th, 1906. The earthquake, which lasted only 67 seconds, was caused by horizontal earth movements along the San Andreas fault. During this short space of time about 700 persons were killed and many buildings were wrecked. The great fire that followed did several million pounds' worth of damage and destroyed a large part of the city. There were landslides in the mountains, great fissures were opened in the ground, and fences and railways were offset by as much as 20 feet.

Kansu, China (1920 and 1927). These two earthquakes in the Kansu region of China claimed over 300,000 lives. Much of the damage was caused by landslides. In those days many of the Chinese lived in caves, and these along with villages and towns were destroyed. Once again earthquakes had altered the geography and geology of part of the world as rivers were blocked by the debris.

Sagami Bay, Japan (1923). This disaster, one of the greatest of modern times, resulted in the loss of almost 100,000 lives and millions of pounds' worth of property. The cities of Tokyo, 70 miles away, and Yokohama, 50 miles away, were heavily damaged. A huge tsunami generated by the submarine earthquake did extensive damage along the coast, and outbreaks of fire consumed about 70 per cent of Tokyo and completely destroyed the city of Yokohama.

Hebgen Lake, Montana (1959). On August 17th, 1959, a series of earthquakes struck in Yellowstone National Park on the Montana–Wyoming border. Much damage was done by landslides which buried campers in the Madison River canyon. The Madison River landslide, involving some 40 million cubic yards of rock, formed a natural dam and created a new lake.

Chile (1960). A series of earthquakes along the coast of Chile beginning in May 1960, were among the most destructive to be recorded in this century. So violent were these shocks that most of the buildings in the area were destroyed, trees and telephone poles were snapped like matchsticks, fissures opened in the ground, and soils flowed like liquid. Further damage was inflicted by seismic sea waves. Ranging from 12 to 30 feet in height, these tsunamis destroyed several Chilean villages and drowned hundreds of people. They crossed the Pacific Ocean at speeds of up to 450 miles per hour and destroyed villages in Hawaii and Japan. The original earthquake of May and subsequent shocks which occurred for a period of several months caused millions of pounds' worth of damage to property and killed more than 5,000 people.

Skopje, Yugoslavia (1963). The television age brought this disaster very graphically before us. Suddenly a little-known city became headline news around the world. Relief workers and equipment were rushed to the area to help the 120 thousand who were homeless and to dig out the 1,000 dead. The water supplies were destroyed and there was a subsequent outbreak of disease.

Alaska (1964). One of the most severe earthquakes ever to affect North America occurred in the vicinity of Anchorage, Alaska, on the afternoon of March 27th, 1964. This earthquake did millions of pounds' worth of property damage, and in places the surface rocks were displaced upward by more than 30 feet. In addition to damage produced by the earth tremors, seismic sea waves wiped out many man-made structures along shorelines in the area. Because the region affected is sparsely populated, only 115 lives were lost—a very small number considering the amount of energy released by this earthquake.

(5) Detecting and Recording Earthquakes

Earthquake waves are detected and recorded by means of a **seismograph** (Fig. 99). Basically, the seismograph consists of a spring-suspended weight or pendulum, free to swing in the direction of the waves to be recorded, and a recording device operated by clock-work. The pendulum is attached to a frame which is embedded in bedrock. Because of its inertia, the free-swinging pendulum is not affected by vibrations in the bedrock; but the rest of the instrument, which is rigidly attached, will be moved by such vibrations. A pen connected to the pendulum records the vibrations on a revolving drum which carries the recording paper. (Some seismographs record on photographic paper which is exposed by means of a spot of light reflected from a mirror.) The resulting records, called **seismograms,** show both the duration and severity of the shock. When the earth's crust is at rest the record will appear as a straight line; earthquake waves will activate the pendulum, producing a wavy line (Fig. 100).

Locating Earthquakes. If satisfactory seismograms have been obtained during an earthquake, these can be used to ascertain where the disturbance occurred. The seismologist wishes to know, for example, the **focus** of the earthquake (the point within the earth from which the shocks originated) and the location of the **epicentre** (the point on the earth's surface directly above the focus). He determines this by making a comparative study of the behaviour of the various types of earthquake waves.

When an earthquake occurs, seismic waves spread out in all directions from the focus, decreasing in intensity with distance. These waves, which vary greatly in amplitude (size) and velocity (speed), are of three basic types. **Primary waves**, also called **P waves**, are

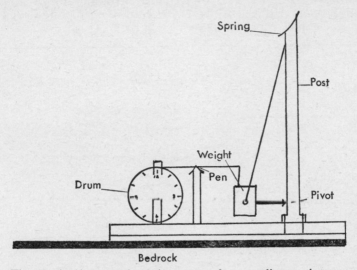

Fig. 99. A seismograph, an instrument for recording earth tremors

compressional waves that travel through the earth at speeds of from 3·4 to 8·6 miles per second. P waves move faster at depth and are the first to be detected by the seismograph. S waves, or **secondary waves**, travel through the earth's interior at speeds of from 2·2 to

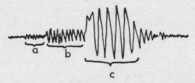

Fig. 100. A seismogram, a recording made by a seismograph: (*a*) primary or P waves, (*b*) secondary or S waves, (*c*) long or L waves

4·5 miles per second. These, the second set of waves to arrive at the seismograph station, will not pass through gases or liquids. **Long waves,** or **L waves,** are complex waves of considerable amplitude which travel near the earth's surface. They originate at the epicentre and are generated from energy produced by P and S waves. L waves which travel relatively slowly (about 2·2 miles per second), are the last to be recorded, and cause most of the earthquake damage.

Fig. 103 at the end of this chapter (after a discussion of the earth's interior) shows the passage of these wave types through the earth.

A study of the relative arrival times of the various types of waves at a single station can be used to determine the distance to the epicentre. If records from at least three widely separated seismograph stations are available, the seismologist can determine the exact location of the epicentre. To do this, circles are drawn on a map, with each of the three stations as the centre of a circle. The point at which the three circles intersect is the epicentre (Fig. 101).

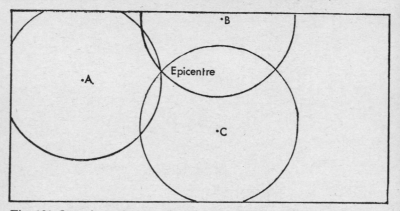

Fig. 101. Locating epicentre of earthquake by using seismograph records from at least three stations

(6) Size of Earthquakes

The size of an earthquake is usually measured in terms of intensity and magnitude.

Intensity. The intensity of an earthquake is measured by the amount of physical damage and/or geologic change it brings about. The shock is most intense at the epicentre, which, as noted earlier, is located on the surface directly above the focus. Damage decreases as distance from the epicentre increases.

Several scales have been devised to indicate varying degrees of earthquake intensity. The so-called **modified Mercalli**, or **Wood-Neumann scale,** is commonly used. This scale uses a series of numbers to indicate different degrees of intensity. These range from an intensity of I: so slight as to be detected only by instruments, to XII: catastrophic earthquakes capable of total destruction. (Originally there were only ten degrees.) When the epicentre is known, the intensity of an earthquake can be indicated on a map by means of **isoseismal lines**—lines connecting areas of equal earthquake intensity. Thus an earthquake of intensity XII at the epicentre will have roughly circular belts of intensity surrounding it, the intensity falling off inversely as the square of the distance from the epicentre.

Magnitude. Because the above scale is subjective and based on the impressions of various people, some seismologists prefer a more quantitative system that relies on instrumental records. These can be used to determine the earthquake's magnitude by assessing the total energy released by the earthquake. This is expressed on the **Richter scale**, a system which indicates earthquake magnitude by means of numbers related to actual energy released in the bedrock.

(7) Interior of the Earth

Although the seismograph's most dramatic work has been in reporting earthquakes, it has also been an important source of

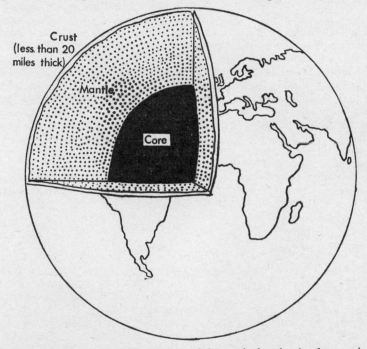

Fig. 102. Diagrammatic cross-section of the earth showing its three major zones

information about the interior of the earth. Data obtained from seismograms indicate that the lithosphere (see Chapter One) may be divided into three zones: the **crust, mantle,** and **core** (Fig. 102).

Crust. The outermost and thinnest layer of the lithosphere is called the crust. The thickness of the crust varies greatly between the ocean basins (as little as four miles in places) and continents (possibly 20–30 miles thick under certain mountains). The specific gravity of the crust ranges from 2·5 to 3·4. Crustal rocks vary not only in thickness and density, but also in composition. Those beneath the ocean basins are heavier than those which underlie the continents. They have been called **sima** because they are rich in iron, *si*licon, and *ma*gnesium. The rocks are primarily of the basaltic type.

The material comprising the continental crust appears to occur in two distinct layers. The upper layer is essentially granitic in nature. Because these rocks contain a high percentage of silicon and aluminium, they are often referred to as **sial.** Evidence derived from the velocities of S and P waves suggests that this sialic layer is 10–15 miles thick. The lower layer, also about 10–15 miles thick, appears to be composed of simatic rocks similar to those underlying the ocean basins.

The base of the crust is marked by a rather clearly defined break called the **Mohorovicic discontinuity** or the **moho.** This sharp boundary, first noted in 1909 by Andrija Mohorovicic, a Yugoslav seismologist, lies 20–30 miles beneath the surface. Here the velocities of S and P waves are somewhat accelerated, implying a change in density in the rocks below the moho.

Mantle. Beneath the Mohorovicic discontinuity there is an 1,800-mile thick intermediate zone called the mantle. The velocities of S and P waves increase gradually upon entering this zone; their behaviour suggests that the mantle is essentially solid and increases in density with depth. The specific gravity of the rocks in this zone ranges from 3·5 (in the upper part of the mantle) to as much as 8·0 at the bottom.

Core. The core of the earth, which is about 4,300 miles (6,800 km) in diameter, is very hot, dense, and under tremendous pressure. It

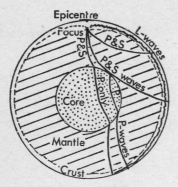

has been divided into two parts: an exterior, probably liquid, outer core, and an inner core which is believed to be solid. The outer core begins at the base of the mantle (about 1,800 miles down) and reaches a depth of about 3,160 miles. The outer core is believed to be fluid because the material in this zone will not transmit S waves, and P waves travel at reduced velocity. The onter core is believed to be about 1,300 miles thick; the materials composing this zone have specific gravities of 12·0 or more.

Fig. 103. Path of shock waves through the earth

The inner core, with a diameter of approximately 1,700 miles, is believed to be solid. This is suggested by the fact that there is an abrupt increase in the speed of P waves deep within the core. It has also been suggested that the inner core may be composed largely of nickel and iron. These rocks are presumed to be quite heavy; some may have a specific gravity of more than 17.

CHAPTER TWELVE

PLAINS, PLATEAUS AND MOUNTAINS

The earth's major land forms, the continents and ocean basins, are said to be relief features of the first order. These have been discussed in earlier chapters. We shall now direct our attention to relief features of the second order: plains, plateaus and mountains.

(1) Plains

Plains, like plateaus (which are discussed below), are underlain by flat-lying, layered rocks. Plains and plateaus differ, however, in their relative elevation above sea level and in the amount of relief present. Most, but not all, plains are relatively near sea level; their relief is seldom more than a few hundred feet.

The different types of plains are usually classified according to the origin of the rock materials which form them.

Marine or Coastal Plains. Interior marine plains, such as those of the upper Mississippi Valley, have been created by uplift with little or no folding or warping. Coastal plains, such as the Atlantic Coastal Plain of the United States, have been formed as a result of emergence of shallow sea floors.

Lake Plains. Known also as **lacustrine** plains, lake plains have been formed by the emergence of a lake floor. Exposure of a lake floor may come about as a result of evaporation, uplift or, more commonly, drainage. Plains of this type are common in North America and Australia.

Alluvial Plains. River plains, or alluvial plains, may be formed as flood plains in river valleys; delta plains, at the mouths of rivers; or piedmont alluvial plains—a series of alluvial fans (see Chapter Seven) formed at the foot of mountains. (Alluvial means composed of mud, sand or other deposits.) Much of Holland is such a plain formed by the Rhine.

Glacial Plains. Glaciers may produce plains in two ways. In some areas, the surface of the underlying horizontal rocks has been levelled by glacial erosion. In other places, outwash plains (see Chapter Eight) are deposited in front of glaciers.

Lava Plains. Plains formed by widely spreading lava flows, either from quiet volcanoes or from great fissure flows, are called lava plains. Good examples of this type of plain are found in Iceland and Hawaii.

(2) Some British Plains

Salisbury Plain. This very well known plain (referred to by some as a plateau) is a large upland area of chalk surrounding the Hampshire Basin.

Irish Plain. The plain which occupies much of central Ireland is really a low-lying plateau. Underlain by limestone, glaciated and suffering from a very humid climate, this part of Ireland is extremely boggy. The Irish Plain may be a late development from post-glacial lakes.

Midland Plain. Southern Lincolnshire, around the Wash, is a silted-up area produced by changes in the courses of rivers such as the Ouse and Welland.

Cheshire Plain. The Cheshire Plain is underlain by a great salt-filled syncline of the Triassic. It is characterized by meres and flashes, some of which are due to 'slumping' following the extraction of the salt.

(3) Plateaus

Plateaus are large, essentially level areas of considerable elevation which are underlain by horizontal rock strata. Unlike plains, plateaus are regions of high relief; their surfaces are normally trenched with ravines and gorges. Most plateaus are more than 2,000 feet above sea level; some, for example the Colorado Plateau and the Tibetan Plateau, are more than a mile above sea level.

Fault Plateaus. In some areas, former plains have been subjected to continuous vertical faulting, which has raised them well above sea level. The fault plateaus thus formed consist of a series of high, nearly horizontal, fault blocks. The Colorado Plateau and the largest plateau of all, the East African Plateau, split by rift valleys, are such fault plateaus.

Warped Plateaus. Some plateaus have been raised by slow uplift, accompanied by little or no faulting. The Appalachian Plateau of the eastern United States is an example of a plateau formed in this manner.

Lava Plateaus. Lava plateaus may be formed when successive horizontal lava flows accumulate to produce a region of high elevation. A large lava plateau extends from Antrim in Ireland to parts of Iceland and Greenland. The largest are the Deccan Plateau of southern India and the Columbia River Plateau of the north-western United States, which both extend over 200,000 square miles and rise several thousand feet.

(4) Mountains

Mountains are regions of considerable relief and high elevation; they have a small summit area and rise conspicuously above the surrounding country. Some geologists restrict the term 'mountains' to those areas in which the rocks have been disturbed or deformed. This would exclude the so-called 'erosional' mountains which have been formed on highly dissected plateaus. Mountains grouped in a series of related ridges to form a continuous unit form a mountain **range.** A mountain **system** is a group of mountain ranges with a common geologic history. A mountain **chain** is an elongated unit composed of several mountain systems, regardless of similarity of form or age relationships.

(5) Origins of Mountains

Mountains may originate as a result of igneous activity (either deep-seated intrusions or volcanic action) or tectonism. They have been classified according to the type of force which formed them.

Volcanic Mountains. Mountains which have been created as a result of extrusive igneous activity are called volcanic mountains (Fig. 104a). These may consist of volcanic necks (Castle Rock at Stirling, Scotland); conical layers of fragmented igneous material around a central vent (Mayon, Philippine Islands); accumulations of lava flows about a central vent (Eildon Hills, Melrose, Scotland); or volcanic domes. Examples of all four types are found in the Hawaiian Islands.

Some of the world's greatest and best-known mountains owe their

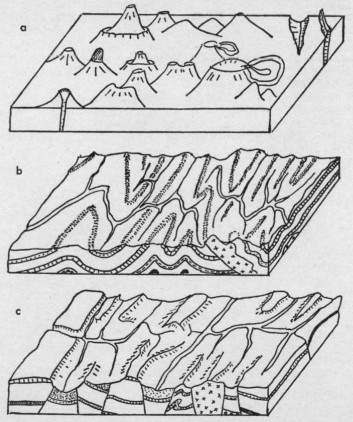

Fig. 104. Types of mountains (*a*) volcanic, (*b*) folded, (*c*) fault or block

F

origin to volcanic activity. These include Etna and Vesuvius in Italy, Fujiyama in Japan, and Popocatepetl in Mexico. In addition, the Hawaiian and Aleutian islands are the tops of volcanic mountain ranges which rise from the ocean floor.

Folded Mountains. Crustal disturbances may cause rock strata to become tightly folded and bent upward for thousands of feet (Fig. 104*b*). Folding, which normally results from compression of the rock strata, may be accompanied by **faulting.** Mountain ranges thus formed consist of alternating upfolds (anticlines) and downfolds (synclines). The Jura Mountains of France and Switzerland are a classic example of folded mountains. The Appalachian Mountains of the eastern United States are a good example of mountains that were created by folding accompanied by faulting. Mountains of this type are sometimes called 'complexly folded' mountains. Other folded mountains include the Alps, Himalayas, Andes and the Rocky Mountains. (Some of the theories advanced to explain the origin of the compressive forces of folding are discussed in Chapter Five.)

Rock strata may be uplifted by igneous intrusions to form broad **domes.** This kind of mountain may be formed when molten rock material invades the bedrock, forcing the overlying strata upward. The type of igneous intrusion most commonly responsible for doming is a **laccolith** (see page 45). The Henry Mountains of Utah are considered to be laccolithic domes. However, not all domal mountains are associated with laccoliths. Some, like the Black Hills of South Dakota, have a broad domal structure and a granitic core, but are not believed to be laccolithic in origin. Although hardly a mountain, Corndon in Shropshire is an excellent example of a dome-shaped hill created in this way.

Fault or Block Mountains. Faulting may cause large blocks of the earth's crust to be lifted upward and tilted at various angles (Fig. 104*c*). The block-faulted mountains thus formed have a short steep slope on one side and a long, more gentle slope on the other. The Vosges and Black Forest mountains are **horsts,** fault blocks which have not been tilted. Part of the Pennines is formed of tilt-blocks dipping gently towards the North Sea.

Complex Mountains. Many of the earth's well-known mountain ranges have been created by a combination of igneous activity and tectonism. Because of their complicated geologic history, they have been called complex mountains. They may show evidence of folding, faulting, volcanic activity, igneous intrusions and doming. Some complex mountains consist almost wholly of igneous rocks, others are composed of metamorphic rocks or gently deformed sedimentary rocks. The mountainous regions of Scotland are complex, folded and faulted in bewildering confusion.

(6) Erosional Remnants

Certain topographic features which do 'rise conspicuously above the surrounding country', but are not composed of disturbed or deformed rocks, are classified as erosional mountains or residual mountains because they are remnants of highlands that have undergone continuous and prolonged erosion.

Features of this sort commonly develop on high, deeply dissected plateaus. They include **mesas** (Fig. 105*a*), broad flat-topped hills, and **buttes** (Fig. 105*b*), smaller steep-sided hills, with narrow tops.

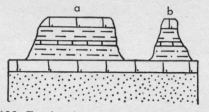

Fig. 105. Erosional remnants: (*a*) mesa, (*b*) butte

Mesas are quite common in Britain, notably in the Midlands and around the Cotswolds. The most famous mesa is Table Mountain near Cape Town, South Africa; buttes, too, are common in South Africa, where they are known as **kopjes.**

When the upstanding beds are tilted we have **cuestas** or **hog's backs** (Fig. 106). Cuestas are common in regions where limestone

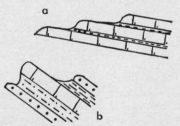

Fig. 106. Erosional features: (*a*) cuesta, (*b*) hog's back

alternates with shale, as in the moors of the Peak District. The famous Hog's Back in Surrey between Farnham and Guildford is formed of a chalk ridge between softer clayey and sandy beds.

Certain **monadnocks** (isolated erosional remnant rocks left standing above a peneplain) are also remains of old mountain ranges. These are fairly common in Australia—Ayers Rock is an example— but they get their name from Mount Monadnock, a residual mountain in New Hampshire, U.S.A.

CHAPTER THIRTEEN

GEOLOGY AND MAN

Man utilizes geologic information and geologic products in an infinite number of ways. Indeed, modern industrial economy depends largely upon the utilization of earth materials of various types. It is the task of the geologist to provide modern civilization with the mineral fuels, ores, and other economic minerals so vital to industrial growth. In this chapter we shall learn something of the nature and importance of some of the more valuable economic products of the earth, how geologic knowledge is utilized by mining and petroleum geologists and engineers, and the way man has altered the face of his planet.

Geology achieves one of its most important aims in the exploration, development, and conservation of our natural **mineral resources.** These include the fossil fuels, metallic minerals, and non-metallic or industrial rocks and minerals.

(1) Fossil Fuels

The fossil fuels, coal and petroleum, are among the most valuable and necessary products of modern industry. The 'Fossil' part of their name reminds us that they are derived from the remains of past life.

Coal. A fossil fuel of plant origin, coal occurs in certain types of sedimentary rocks. It consists largely of carbon, hydrogen, oxygen, and nitrogen, but usually has a certain amount of sulphur, silica, and aluminium oxide as impurities. Coal is formed by **carbonization,** a process by which decaying plant material loses water and volatile substances with a resulting concentration of carbon (see also Chapter Fifteen). The various kinds of coal and their method of formation were discussed in Chapter Four.

Coal is not only plentiful, it is widely distributed over the earth. Germany, the United States and Great Britain are the major coal-producing countries of the world. Most of our coal originates from the remains of Carboniferous plants. In Britain coal is mined mainly in the Midland Valley of Scotland, Northumberland, Durham, Cumberland, Lancashire, Yorkshire, Derbyshire, Nottingham, Leicestershire, Stafford, Shropshire, North and South Wales and Kent. Although petroleum has replaced coal in many phases of industry, around 150 million tons of coal are produced in Britain in a year.

Petroleum. Most geologists believe that petroleum (oil and gas) originated from the remains of microscopic marine plants and animals. These remains, which were buried in the mud and sand of shallow prehistoric seas, underwent slow decomposition by bacteria, leaving a residue of hydrocarbon compounds. Although the processes by which the organic material was finally converted to petroleum are not thoroughly understood, they appear to have required vast amounts of time, accompanied by increased temperature and compression of the sediments.

After its formation, the petroleum moved from the muds and shales in which it was formed into more porous rock. It then migrated to rock structures favourable for petroleum concentration. The sediments in which the oil and gas were formed are called **source rocks.** These are typically dark clays and shales of high organic content. The porous and permeable rocks through which the oil migrates are known as **reservoir rocks.** Sands, sandstones, porous limestones and dolomites make effective reservoir rocks. Structures (or **traps**) are areas in the reservoir rock which will stop oil and gas migration and cause it to accumulate. Anticlines, faults and salt domes are all effective traps. We see, then, that in order to have an **oil pool** (a porous bed or rock saturated with oil) we must have (*a*) a source bed, (*b*) reservoir rocks and (*c*) traps or structures.

The main task of the petroleum geologist is to locate traps that are suitable for stopping the migration of oil and gas. This search is carried on in a number of ways: the geologist may study and map rocks which are exposed at the surface, or he may examine rock fragments brought to the surface by exploratory drilling. In addition, many oil companies employ geophysical methods in their search for oil. This kind of exploration requires a type of seismograph similar to that used to record earthquakes (see Fig. 99). The technique, known as geophysical prospecting, involves the production of small, artificial 'earthquakes' by means of explosives. The seismograph records the path of the shock waves as they travel through the rocks; seismic records secured in this manner give some indication as to the type of rocks present, their relative depth, and whether or not a suitable trap may be present.

Petroleum is found in many parts of the world and in rocks ranging in age from Cambrian to Late Tertiary. In some areas it is produced from a few feet beneath the surface; in others it must be taken from rocks several miles deep. Oil production in the United States (where Texas is the leading producer) greatly exceeds that of other countries. The first sale of 'lots' in Alaska for oil prospecting took place in September 1969. Astronomical sums were bid in a strange auction bringing in billions of dollars. The Soviet Union, Romania, the Middle East (Saudi Arabia, Persia and Iraq), Indonesia, Venezuela, Colombia and other South American countries, Canada and Mexico are some of the more important oil-producing areas. There are a few deposits in parts of England. The oilfields in Yorkshire have not yet proved economically worth tapping, but a small well to the north-east of Nottingham does produce tens of thousands of tons of oil per year. More important than these has been the find of natural gas in the North Sea, and gas appliances are being rapidly converted to use this valuable product. Recently oil and natural gas have been discovered in Australia and are supplying Brisbane, and in March 1970 came news of a further Australian strike.

(2) Metallic Minerals

The metallic or ore minerals include such valuable substances as aluminium, copper, gold, lead, mercury, silver, tin, zinc, iron and nickel. Important also are the radioactive minerals such as uraninite (or pitchblende) and carnotite. The occurrence, use, and physical and chemical characteristics of the more important metallic minerals have been discussed in Chapter Two.

Metallic minerals may occur in igneous, sedimentary, or metamorphic rocks. The ores of many metals commonly occur in **veins.** Some veins are formed when circulating ground water picks up metallic compounds and deposits them in crevices in the rock. Others, associated with igneous activity, occur as a result of magma being injected into the country rock. The latter deposits are commonly associated with contact metamorphic zones (see Chapter Five) along the periphery of igneous intrusions.

In some areas there are residual concentrations of ores which are the product of chemical weathering. Thus, the valuable aluminium ore bauxite may be formed as the result of the weathering of certain clays, granites, or syenites of high aluminium content. Some of the world's larger deposits of iron ore are believed to have been formed in a similar manner.

Metallic minerals also occur in natural mechanical concentrations called **placer deposits.** Ore accumulations of this type have been found in sands and gravels in the beds of streams that eroded the rocks in which the metals were originally formed. Gold from the bed of the Sacramento River, California, and (with platinum) from gravels in South Africa; tin from gravels in Malaya; precious non-metals, such as diamonds from Africa: these are some of the famous placer deposits being exploited today.

Some ore deposits appear to be the result of deposition of minerals in prehistoric lakes and seas. These include some of the larger iron-ore deposits of the United States and France, and the enormous manganese deposits in Russia.

Today the well-trained mining geologist employs a host of geophysical and geological techniques and instruments that have been developed to replace the screening-pan and pick of the gold-rush days.

(3) Non-metallic or Industrial Rocks and Minerals

In addition to the fossil fuels and metallic minerals, there is an important group of rocks and minerals that are utilized, not for any metals they may contain, but for some other purpose. Included here are such valuable materials as asbestos, quartz, sand, clay, cement, mineral fertilizers, salt, lime and sulphur.

Building stones such as sandstone, granite, limestone, slate and marble are other non-metals of considerable economic importance. They, like the metallic minerals, have been discussed in earlier chapters.

(4) Engineering Geology

Engineering geology is the application of geology to various problems and procedures in civil engineering. For example, many undertakings such as dams, bridges, canals, reservoirs, tunnels, tall buildings and other heavy structures could not be successfully completed without considering certain basic geologic problems. This consideration has not, unfortunately, always been given; various landslides, dam failures and tunnel collapses may be attributed to faulty engineering practices which neglected to allow for certain geologic conditions.

(5) Man, the Geologic Agent

Earlier chapters have discussed the geologic agents of water, ice, wind, gravity and sea, but they omitted one and very potent agent: man. During the 35,000 years that *Homo sapiens* has walked the earth, he has increasingly altered the world around him.

We have already noted the effect of man's activities in connection with lakes and dams and reservoirs. The reservoirs in Britain, including those in the Welsh mountains which supply some of our industrial needs, are puny compared with the huge artificial lakes being created in Africa and elsewhere. The Snowy Mountains Project in Australia started 25 years ago and now nearly complete, is turning 3,000 square miles of beautiful mountain scenery into a huge lake called Eucumbene, and has altered the courses of three large rivers. A scheme exists in Britain to flood Morecambe Bay and create a much-needed reservoir there.

Quarries have caused great scars in hillsides (building stones from Dartmoor and the Pennines, clays from the London Basin). Opencast mining, such as the Mount Isa complex which provides much of Australia's copper, silver, zinc and lead, disfigures the landscape. Many clay pits are flooded when their economic value is over. Such a quarry existed at Nyewood, near Petersfield, Hampshire, where a small brick-making concern used the Gault clay until the early 1960s. Here were found excellent examples of ammonites, gypsum crystals and phosphatic and iron pyrites nodules. Now it is flooded and a glass-works has been set up on its doorstep. Often new quarries and pits are permitted only if the land can be either reclaimed as a lake or filled in and grassed over.

Slag heaps disfigure the countryside creating man-made hills. Other hills, such as the *tels* in the Middle East, are found over the ruins of ancient towns.

The wind's erosional work has been assisted by man, who has sometimes cut back the natural vegetation and farmed the land only to find that all he has left after a few years is a dust-bowl.

Mining districts are prone to subsidence or slumping (we have already mentioned the 'flashes' and 'meres' that have been created in mid-Cheshire as a result of brine-pumping).

The recent flurry of activity connected with the Alaskan oil finds will change the face of Alaska. Such projects cause concern to the conservationist. Already man has contributed to the extinction of many species of animal life and is threatening more. Now at last, but too late for some, man is learning to safeguard the future of animal life, as well as scenic and other amenities, on his planet.

PART II

HISTORICAL GEOLOGY

CHAPTER FOURTEEN

THE ORIGIN AND AGE OF THE EARTH

The first part of this book was concerned primarily with the physical aspects of the earth and the geologic processes that are working on it. In this and succeeding chapters we shall discuss the origin and age of the earth, and attempt to reconstruct some of the more important events of the geologic past. Finally, we shall see how a geological map records a wealth of physical and historical information for a given area.

(1) The Origin of the Earth

Whence came the earth? How did it all begin? Man has speculated on such questions since the beginning of recorded history. This problem, one that is still unsolved, has resulted in the development of a number of hypotheses—none of which is entirely satisfactory.

The Nebular Hypothesis. This hypothesis suggests that the solar system developed from a **nebula**—a vast disc-shaped cloud of gas. This idea was first proposed in 1755 by the German philosopher Immanuel Kant. Later, in 1796, the French mathematician Pierre Laplace developed the theory more fully and stated it in more scientific terms. Although these men arrived at similar conclusions, Laplace was not aware of Kant's earlier work.

Kant and Laplace assumed that, at some period in the distant past, a great nebula—its diameter reaching beyond the orbit of our outermost planet—was slowly rotating in space. As this gaseous mass cooled, it shrank and rotated more and more rapidly. Eventually the rim of the nebula rotated so fast that centrifugal force overcame gravitational force, and a ring of gas separated itself from the parent body. The nebula continued to contract, and the speed of rotation continued to increase, until a total of ten rings had been thrown off. Nine of these rings slowly condensed to form our planets. The sixth ring, rather than condensing into a single body, broke up into many small masses. These small bodies formed the asteroids, and the central mass of the nebula later condensed to form the sun.

This theory was quite popular and gained much scientific support in the nineteenth century. Unfortunately, later research proved it to be untenable and it was abandoned in the early twentieth century. There are numerous objections to it, but the most important is that this mechanism would not work because the sun rotates too slowly in comparison with the rest of the planets.

The Planetesimal Hypothesis. According to this hypothesis, the sun was originally a star which existed without planets. At some time in the remote past, another star passed very close to the sun, exerting a gravitational force powerful enough to tear great masses of material from opposite sides of the sun. As the matter was pulled out from the sun, it cooled and condensed into solid particles called **planetesimals.** The largest of these planetesimals acted as nuclei which attracted other planetesimals, and by accretion the planets slowly grew to their present size—each pursuing its own orbit around the sun. The satellites were supposed to have formed from small clusters of planetesimals located near the nuclei from which the planets originated.

Although this hypothesis was widely accepted for several decades, a number of geological and astronomical objections have been raised to it. For example, much of what we know of the structure of the earth suggests that it was originally in a molten condition, whereas the hypothesis postulated an originally solid planet. In addition, there is some doubt that the planetesimals could have gathered together by accretion, for the collision of these particles in outer space would probably have destroyed them.

The Tidal or Gaseous Hypothesis. Like the planetesimal hypothesis which preceded it, this also involves an original sun that had a close encounter with a passing star. It was proposed by Sir James Jeans, an astronomer, and Sir Harold Jeffreys, a geophysicist, in an attempt to counteract some of the objections that had been raised to the planetesimal hypothesis. They accepted the supposed near-collision between the sun and another star, but believed that the material pulled from the sun came out as a long cigar-shaped filament of solar gases. This gaseous filament later broke up into units which condensed to a molten and finally a solid stage, thus forming the planets. Astronomers, however, have shown that a gaseous filament would not form solid bodies such as our planets; it would instead simply disappear in space. For this and other reasons, the hypothesis is no longer acceptable to most scientists.

Recent Advances in Cosmology. New developments in mathematics, physics and astronomy have resulted in much new speculation about the origin of our solar system. Some of the newer ideas are the Electromagnetic Hypothesis of Alfven (1942), the Nebular-Cloud Hypothesis of von Weizsäcker (1944), the Nova Hypothesis of Hoyle (1945) and the Dust-Cloud Hypothesis of Whipple (1947).

(2) The Age of the Earth

Now that we have speculated as to *how* the earth was formed, we should give some thought as to *when* it was formed. Estimates of the earth's age have ranged from as few as 6,000 years (by early theologians) to as many as 10,000 *million* years (by astronomers and physicists). However, the latest scientific evidence indicates that the earth is probably closer to 4,500 million years old.

How do we know? Before attempting to answer this question we will turn our attention to the geologic time scale for this will help us to understand the great antiquity of our planet.

(3) The Geologic Column and the Geologic Time-Scale

The geologic column refers to the total succession of rocks, from the oldest to the most recent, that are found in the entire earth or in a given area. Thus the geologic column of a given region would include all rock divisions known to be present in that region. By referring to the geologic column previously determined for a specific area, the geologist knows what kind of rocks he might expect to find in that particular region.

The geologic time-scale (Fig. 107) is composed of named intervals of geologic time during which the rocks represented in the geologic column were deposited. These intervals of time bear the same names that were originally used to distinguish the rock units in the column. For example, one can speak of the Ordovician period (referring to the geologic time-scale) or of Ordovician rocks (referring to the geologic column).

Both the geologic column and the geologic time-scale are based on the **principle of superposition.** This obvious but important principle states that, unless a series of sedimentary rock has been over-turned, a given rock will be older than all the layers above it and younger than all the layers below it. The field relationship of the rocks plus the type of fossils (if present) gives the geologist some indication of the relative age of the rocks. Relative age does not indicate age in years; rather, it fixes age in relationship to other events that are recorded in the rocks.

Recently, however, it has become possible to assign ages in years to certain rock units. This system of rock dating, based largely on the presence of radioactive minerals in the rocks (described later in this chapter), has made it possible to devise an absolute time-scale which gives us some idea of the tremendous amount of time that has passed since the oldest known rocks were formed. It has also been used to verify the previously determined relative ages of the various rock units.

Units of the Time-Scale. The largest unit of geologic time is an **era,** and each era is divided into smaller time units called **periods.** A period of geologic time is divided into **epochs,** which in turn may be subdivided into **series** and **zones.** The geologic time-scale might be roughly compared to the calendar, in which the year is divided into months, months into weeks, and weeks into days. It should be emphasized, however, that, unlike years, geologic time units are arbitrary and of unequal duration; thus the geologist, dealing with relative time, cannot be positive about the exact amount of time involved in each unit. The time-scale does, however, provide a standard by which one can discuss the relative age of the rocks and the fossils they contain. For instance, by referring to the time-scale, it is possible to state that a certain event occurred in the Palaeozoic Era in the same sense that one might say that something happened during the Reformation.

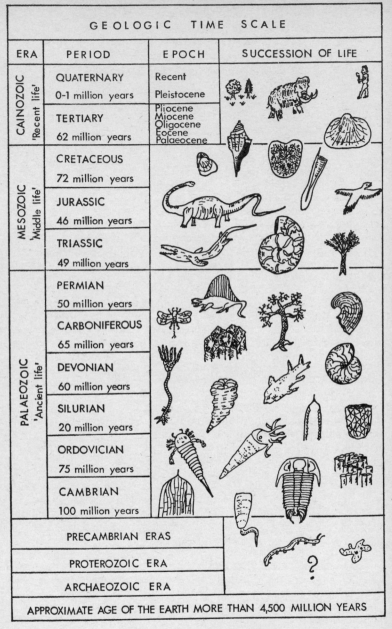

ERA	PERIOD	EPOCH	SUCCESSION OF LIFE
CAINOZOIC 'Recent life'	QUATERNARY 0-1 million years	Recent / Pleistocene	
	TERTIARY 62 million years	Pliocene Miocene Oligocene Eocene Palaeocene	
MESOZOIC 'Middle life'	CRETACEOUS 72 million years		
	JURASSIC 46 million years		
	TRIASSIC 49 million years		
PALAEOZOIC 'Ancient life'	PERMIAN 50 million years		
	CARBONIFEROUS 65 million years		
	DEVONIAN 60 million years		
	SILURIAN 20 million years		
	ORDOVICIAN 75 million years		
	CAMBRIAN 100 million years		
PRECAMBRIAN ERAS			
PROTEROZOIC ERA			
ARCHAEOZOIC ERA			
APPROXIMATE AGE OF THE EARTH MORE THAN 4,500 MILLION YEARS			

GEOLOGIC TIME SCALE

Fig. 107. Geologic time-scale

There are five eras of geologic time, and each has been given a name that is descriptive of the degree of life development appropriate to it. Palaeozoic, for example, means 'ancient-life', and the era was so named because of the relatively simple and ancient stage of life development of that time.

The five eras and the literal translation of each name are shown below.

Cainozoic—recent life
Mesozoic—middle-life
Palaeozoic—ancient-life
Proterozoic—fore-life
Archaeozoic—beginning-life

The oldest era is at the bottom of the list because this part of geologic time transpired first and was then followed by the successively younger eras which are placed above it. Therefore, the geologic time-scale is always read from the bottom of the chart upward.

Archaeozoic and Proterozoic rocks are commonly grouped together and referred to as **Pre-Cambrian.** The record of this part of earth history contains very few fossils, and is most difficult to interpret. It has been estimated that Pre-Cambrian time may represent as much as 85 per cent of all geologic time.

The Palaeozoic, Mesozoic and Cainozoic eras have each been divided into periods, and most of these periods derive their names from the regions in which the rocks of each were first studied. For example, rocks of the Devonian periods were first studied in Devonshire. (The succession of rocks in these source regions are known as **type sections.**)

The Palaeozoic Era has been divided into six periods of geologic time. With the oldest at the bottom of the list, these periods and the sources of the names are:

Permian—from the province of Perm in Russia
Carboniferous—from its association with coal-forming vegetation
Devonian—from Devonshire
Silurian—from the Silures, an ancient tribe of Wales
Ordovician—from the Ordovices, an ancient tribe of Wales
Cambrian—from *Cambria*, the Latin word for Wales
(American geologists tend to substitute two periods, the Mississippian and the Pennsylvanian, for the Carboniferous period.)

The periods of the Mesozoic Era and the sources of their names are:

Cretaceous—from the Latin word *creta*, meaning 'chalky'
Jurassic—from the Jura Mountains of France and Switzerland
Triassic—from the Greek word *trias*, meaning 'three'
The Cainozoic Era is divided into two periods:

Quaternary
Tertiary

The names of these two periods are all that remains of an outdated system of classification in which all rocks were divided into four groups.

Units of Rocks. A geologist works with small units of rock to build up the whole picture of a district or an era. A **bed** is an individual layer of rock. Beds of rock build up to give **zones** named from a characteristic fossil; for example, the highest chalk zone in Britain is named the Mucronata zone after a belemnite. The chalk is but one **formation** found in the Cretaceous **system**. Intermediate between the zone and the formation is the stage, for example the Danian stage (though this is not found in Britain). Several formations give a **series**; the Upper Cretaceous includes the Gault, Upper Greensand and Chalk formations.

Often a formation is given a double name which indicates both its location and rock type. For example, the Wenlock Limestone (Silurian) is a limestone which forms Wenlock Edge in Shropshire.

(4) Measuring Geologic Time

Geologists have used a number of methods to estimate the age of the earth. Some of these have proved useless, or given at best only a very rough approximation. Others (for example, certain of the radioactive methods) are much more accurate.

Salinity of the Sea. The oceans were probably originally composed of fresh water; they have become saline as salt has been dissolved from the soil and added to the sea by rivers. Age estimates based on this method suggest that it would have taken about 100 million years for the seas to attain their present degree of salinity. The fact that much of the salt has gone through several cycles of deposition, and numerous other problems associated with this method, have made it unworkable.

Rate of Sedimentation. It has been suggested that if it were known how long it took to deposit all of the rock layers in the crust, we could get some idea as to the age of the earth. This is done by measuring the thickness of the strata and multiplying these figures by the rate at which they are assumed to have been deposited. Estimates derived in this way range from 100 to 600 million years. Because of greatly varying rates of deposition and erosion, this method has little scientific value.

Radioactive Methods. These are the most recent and accurate methods yet devised. Certain radioactive elements, notably uranium, undergo slow spontaneous disintegration. The rate of disintegration is constant, and is not affected by changes in temperature, pressure, or other natural conditions. Helium is released as the mineral disintegrates, and a new series of elements is formed. The last element formed in this series is lead. By calculating the ratio between the amount of lead and the remaining amount of uranium present in a given specimen, it is possible to determine the age of the radioactive mineral. This method is limited, of course, to those rocks containing radioactive minerals. Similar methods based on the rate of decay of rubidium to strontium, and potassium to argon, are also being used with some accuracy. The oldest rocks dated by radioactive methods indicate that the earth is approximately 4,500 million years old.

The **Carbon-14** method of dating has proved successful in dating objects less than 40,000 years old. Briefly, this method is based on the fact that all organisms contain a constant amount of carbon-14, a radioactive form of carbon which has an atomic weight of 14 instead of the usual 12. When an organism dies, the carbon-14 is gradually lost by radioactive disintegration (or decay) which proceeds at a known rate. This rate is such that half of the carbon-14 has disintegrated at the end of 5,568 years. The approximate age of a specimen may be determined by comparing the amount of carbon-14 remaining in it with the amount present in most living things. This method has been extremely useful in dating both geological and archaeological materials.

CHAPTER FIFTEEN

THE RECORD OF THE ROCKS

How does the historical geologist know what happened to this planet millions—or even thousands of millions—of years ago? He learns by studying the record of the rocks, a record which indicates that both the earth and its inhabitants have undergone many changes throughout the long life span of our planet. What does this record consist of? By what means does the geologist interpret it? These are some of the questions that will be considered in this chapter.

(1) Keys to the Past

We now believe that the earth is at least 4,500 million years old. There is, moreover, evidence which suggests that life may have been present on our planet for as long as 3,000 million years. In addition, there are many indications that the earth's physical features have not always been as they are today. For instance, mountains now occupy the sites of ancient seas, and coal is being mined where swamps existed millions of years ago. Furthermore, plants and animals have undergone great change. The trend of this organic change is, in general, toward more complex and advanced forms of life. However, some life forms have remained virtually unchanged and others have become extinct.

In order to interpret earth history, the geologist must study the rocks and gather evidence of the great changes in geography, climate and life that took place in the geologic past. To do this he has developed several methods or principles which act as keys to the more important events of prehistoric time.

The Doctrine of Uniformitarianism. This important geologic principle states that the geologic processes of the past operated in essentially the same manner and at the same rate as they do today. In other words, *the present is the key to the past*. This means, in effect, that the earth features of today have been formed as the result of present processes acting over long periods of time.

The Law of Superposition. In an undisturbed sequence of sedimentary rocks, the rocks at the bottom of the sequence are older than any of the rocks which overlie them (Fig. 108). This fundamental geologic concept, the law of superposition, is basic to an understanding of geologic history. Where the rocks have been greatly disturbed, it is, of course, necessary to determine the tops and bottoms of beds before the normal sequence can be established.

Relative Age of Igneous Rocks. Extrusive rocks, such as lava flows, are obviously younger than the rocks on which they rest. Intrusive rocks, for example dykes, sills and batholiths, are younger than the rocks into which they have been injected. Thus the dyke labelled 2 in Fig. 108 is younger than Bed 1 but older than Bed 3.

The Law of Faunal Succession. This law states that fossil faunas (assemblages of animals that lived together at a given time and place) follow each other in a definite and determinable order. These faunas are distinctive for each portion of earth history, and by comparing them the geologist is able to recognize deposits of the same age. This succession of life is such that older rocks may be expected

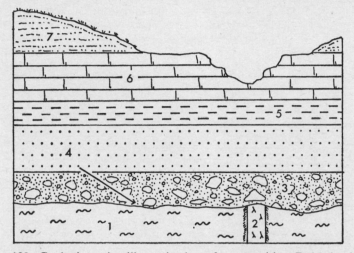

Fig. 108. Geologic section illustrating law of superposition. Bed 1, eroded metamorphic rock, is the oldest; an unconformity (arrowed) separates it from Bed 3. The beds become progressively younger as they ascend; Bed 7 is the youngest

to yield the remains of more primitive organisms, while the remains of more advanced life forms are normally confined to the younger rocks.

Correlation. Correlation, the process of determining the relative ages of the rocks exposed in different areas (or from rock samples from different drillings), is one of the most important tools of the geologist. It is an especially valuable technique because no single area can provide a rock section containing a record of all geologic time. But since deposition has always been continuous in one place or another we can correlate widely scattered outcrops to compile a composite record of all geologic time.

Although there are numerous methods of correlation, those most frequently used are:

Continuity of Outcrops. Rock layers that may be traced without interruption are the easiest to correlate. This method is usually limited to use in relatively small areas, however.

Lithologic Similarity. Some formations are relatively consistent in their rock characters, and this uniformity may be used in tracing the formation from one area to another. This method should be used with caution, however, as many rock units undergo some change in texture or composition from one outcrop to another. On the other hand, some formations are marked by the presence of such diagnostic features as unusual weathering patterns, distinctive mineral associations, peculiar concretions, and the like.

Similarity of Sequence. The geologist often correlates by comparing the positions in which certain strata appear in widely separated vertical sections. If, for example, a red sandstone is known to occur normally between a very coarse conglomerate and black shale, this

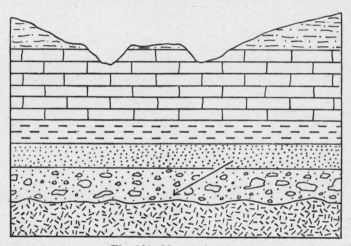

Fig. 109. Nonconformity

sequence will be relatively easy to recognize in the field. Unconformities (see below), if present, may be used in a similar manner.

Similarity of Fossils. If fossils are present in the rocks, they are especially useful for purposes of correlation. Those fossils having a limited vertical range but a rather wide geographic distribution are particularly useful and are called **guide** or **index** fossils. Correlation by means of fossils is discussed in some detail later in this chapter.

Unconformities. At many places in the geologic record there is evidence of crustal uplift followed by long periods of erosion or non-deposition. Such a break or gap in the record is called an unconformity. Geologists recognize three basic types of unconformities.

Nonconformity. This type of unconformity is formed when overlying stratified rocks lie on an eroded surface of igneous rocks (Fig. 109).

Disconformity. In unconformities of this type, the rock layers above and below the unconformity are parallel (Fig. 110).

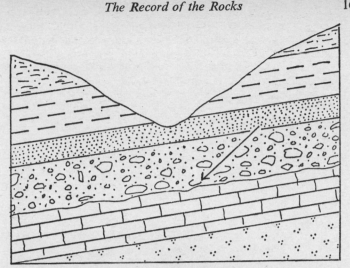

Fig. 110. Disconformity. (Note parallel beds above and below the unconformity)

Angular Unconformity. In this, a relatively obvious type of unconformity, the beds above the unconformity are not parallel to the beds below it (Fig. 111). This type of unconformity indicates that the lower series of rocks were tilted or folded prior to their erosion and the subsequent deposition of the overlying beds.

Palaeogeography. This branch of historical geology is concerned with the distribution and relationships of ancient seas and land masses. Ancient geography (which is what palaeogeography literally

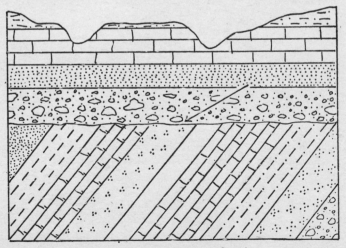

Fig. 111. Angular unconformity

means) is reconstructed through an interpretation of the sedimentary rocks of a certain age. Fossils, if present, are also valuable in determining past geographic conditions.

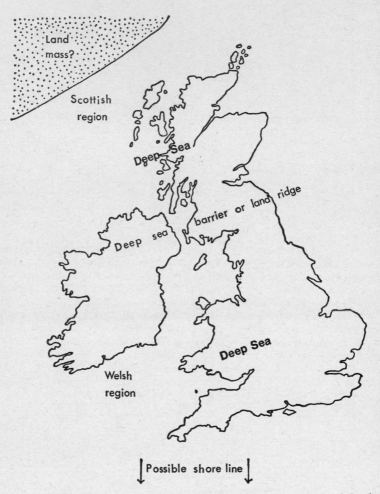

Fig. 112. Simplified palaeogeographic map showing extent of lands (shaded) and seas during the Cambrian Period

If in a given area we find, for example, sedimentary rocks containing marine fossils of Late Cretaceous age, this would indicate the presence of seas in this area in Late Cretaceous time. Restorations of these ancient geographic features are shown on palaeogeographic maps (Fig. 112).

(2) Fossils: The Record of Life on Earth

Palaeontology—the study of fossil plant and animal remains—has done much to advance our knowledge of prehistoric life. Because it is concerned with the record of life, palaeontology is closely related to the science of biology. In studying fossils, the palaeontologist relies heavily on uniformitarianism, the principle that states that 'the present is the key to the past'. Because of the great amount of time involved in earth history, we are not always certain of the life habits or environments of extinct plants and animals. However, when we discover a fossil group whose members closely resemble the members of a group existing today, it is usually safe to infer that the fossil organisms lived under conditions similar to those of the existing group.

(3) The Divisions of Palaeontology

Fossils represent the remains of such a diversity of organisms that palaeontology has been divided into four main divisions.

Palaeobotany deals with the study of fossil plants and the changes which they have undergone.

Invertebrate palaeontology is the study of fossils without a backbone or spinal column. Fossils of this type include protozoans (tiny one-celled animals), trilobites, echinoids, molluscs and brachiopods, and usually represent the remains of animals that lived in prehistoric seas.

Vertebrate palaeontology is the study of ancient animals with a backbone or spinal column. Included here are the remains of fish, amphibians, reptiles, birds and mammals.

Micropalaeontology is the study of fossils so small that they are best examined with a microscope. These small remains, called **microfossils,** usually represent shells or parts of minute plants or animals. Microfossils are especially valuable to the petroleum geologist, as they enable him to identify rock formations that are thousands of feet below the surface.

(4) How Fossils are Formed

Herodotus, in 450 B.C., noticed marine fossils in the Egyptian desert and correctly concluded that the Mediterranean Sea had once extended to that area. During the 'Dark Ages', fossils were alternately explained as freaks of nature, or devices of the devil placed in the rocks to lead men astray. Such superstitious beliefs hindered the development of palaeontology for several centuries. But in the last 100 years, fossils have been accepted without question as the remains of ancient life, and have become increasingly important to the geologist.

The majority of fossils are found in marine sedimentary rocks. Such rocks were formed when salt-water sediments such as lime muds, sands, or shell beds were compressed and cemented together to form rocks. Only rarely do fossils occur in igneous and metamorphic rocks.

But even in the sedimentary rocks only a minute fraction of prehistoric plants and animals have left any record of their existence. This is not difficult to understand if we are aware of the rather rigorous requirements of fossilization.

Requirements of Fossilization. There are many factors which ultimately determine whether an organism will be fossilized, but the three basic requirements are:

(*a*) **The organism should possess hard parts.** These might be shell, bone, teeth or the woody tissue of plants. However, under unusually favourable conditions of preservation, it is possible for even such fragile creatures as a jellyfish or an insect to become fossilized.

(*b*) **The organic remains must escape immediate destruction after death.** If the body parts of an organism are crushed, decayed, badly weathered or otherwise greatly changed, this may result in the alteration or complete destruction of the fossil record of that particular organism.

(*c*) **Rapid burial must take place in a material capable of retarding decomposition.** The type of material burying the remains usually depends upon where the organism lived. The remains of marine animals are common as fossils because they fall to the ocean bottom after death, and here they are covered by soft muds which are converted into the shales and limestones of later geologic ages. The finer sediments are less likely to damage the organic remains, and certain fine-grained Jurassic limestones in Germany have faithfully preserved such delicate specimens as birds, insects and jellyfishes.

Ash falling from nearby volcanoes has been known to cover entire forests (and, as at Pompeii, cities). Some of the resulting fossil forests have been found with the trees still standing and in excellent states of preservation. Good examples of such trees can be seen in Yellowstone National Park, Wyoming. Fossil forests in Britain are found as far apart as Lulworth in Dorset and Glasgow in Scotland, but most of these, such as those in the bays of Swansea and Barnstaple, are preserved by **petrification**—a process described later on.

Quicksand and tar have also been responsible for the rapid burial of animals. The tar acts as a trap to capture the beasts and as an antiseptic to retard the decomposition of their hard parts. The Rancho La Brea tar pits at Los Angeles, California, are famous for the large number of prehistoric animal bones that have been recovered from them. These include the sabre-toothed tiger, giant ground sloths, and other creatures now extinct. The remains of animals that lived during the Ice Ages have been preserved in the ice or frozen ground, and some of these, especially the great woolly mammoths, are still intact to a remarkable degree.

(5) Gaps in the Fossil Record

Although untold numbers of organisms have inhabited this planet in past ages, only a minute fraction of them have left any record of their existence. Even if the basic requirements of fossilization have been fulfilled, there are still other reasons why some fossils may never be found.

For example, large numbers of fossils have been destroyed by erosion, or their hard parts have been dissolved by underground waters. Others were buried in rocks that were later subjected to great physical change, and fossils found in these rocks are usually so damaged as to be unrecognizable.

Then, too, many fossiliferous rocks cannot be studied because they are covered by water or great thicknesses of sediments, and still others are situated in places that are geographically inaccessible. These and many other problems confront the geologist as he attempts to describe the plants and animals of the past.

The gaps in the fossil record become more numerous and more obvious in the older rocks of the earth's crust. This is because the more ancient rocks have had more time to be subjected to physical and chemical change or to be removed by erosion.

(6) The Different Kinds of Fossil Remains

The division of fossils into four types is based upon the composition of the remains or the changes which they have undergone since their burial.

Original Soft Parts of Organisms. For its soft parts to be preserved, an organism must be buried in a medium capable of retarding decomposition. Materials known to have produced this type of fossilization are frozen soil or ice, oil-saturated soils and amber (fossil resin). It is also possible for organic remains to become so **desiccated** (dried out) that a natural mummy is formed. This is likely to occur only in arid or desert regions and in situations where the remains have been protected from predators and scavengers.

The best known examples of preserved soft parts of prehistoric animals have been discovered in Alaska and Siberia. The frozen tundra of these areas has yielded the remains of large numbers of frozen woolly mammoths, an extinct elephant-like animal. The bodies of these huge beasts, many of which have been buried for as long as 25,000 years, are exposed as the frozen earth begins to thaw. Some of these giant carcasses have been so well preserved that their flesh is still edible, and their tusks have been sold by ivory traders.

Original soft parts have also been recovered from oil-saturated soils in eastern Poland. The well-preserved nose-horn, a foreleg, and part of the skin of an extinct rhinoceros were collected from these deposits.

The natural mummies of ground sloths have been found in caves and volcanic craters in New Mexico and Arizona. The extremely dry desert atmosphere permitted thorough dehydration of the soft parts before decay set in, and specimens with portions of the original skin, hair, tendons, and claws have been discovered.

Another interesting and unusual type of fossilization is preservation in amber. This type of preservation was made possible when ancient insects became trapped in the sticky gum that exuded from certain coniferous trees. With the passing of time this resin hardened, leaving the insect encased in a tomb of amber. Some of these insects and spiders have been so well preserved that even fine hairs and muscle tissues may be studied under the microscope.

The preservation of original soft parts has produced some interesting and spectacular fossils. However, this type of fossilization is relatively rare, and the palaeontologist usually finds it necessary to study remains that have been preserved in stone.

Original Hard Parts of Organisms. Almost all plants and animals possess some type of hard parts which arc capable of becoming fossilized. Such hard parts may consist, for example, of the shell material of clams, oysters or snails, the teeth or bones of vertebrates, the exoskeletons (outer body coverings) of crabs or the woody tissue of plants. These hard parts are composed of various materials which are capable of resisting weathering and chemical action, and fossils of this sort are relatively common.

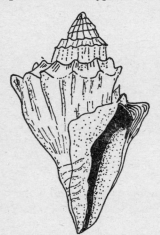

Fig. 113. Fossil shell composed of calcium carbonate

Calcareous Remains. Hard parts composed of calcite (calcium carbonate) are quite common among the invertebrates. This is especially true of the shells of clams, snails, and corals, and many of these shells have been preserved with little or no evidence of physical change (See Fig. 113).

Phosphatic Remains. The bones and teeth of vertebrates and the exoskeletons of many invertebrates contain large amounts of calcium phosphate. This compound is particularly resistant to weathering, hence phosphatic remains are found in an excellent state of preservation.

Siliceous Remains. Many organisms having skeletal elements composed of silica (silicon dioxide) have been preserved with little noticeable change. The siliceous hard parts of many microscopic organisms and certain types of sponges have become fossilized in this manner.

Chitinous Remains. Some animals have an exoskeleton composed of chitin, a material that is similar in composition to our fingernails. The fossilized chitinous exoskeletons of arthropods and other organisms are commonly preserved as thin films of carbon because of their chemical composition and method of burial.

Altered Hard Parts of Organisms. The original hard parts of an organism normally undergo great change after burial. Such changes take place in a variety of ways, but the type of alteration is usually determined by the composition of the hard parts and where the organism lived. Some of the more common processes of alteration are described below.

Carbonization. This process, known also as *distillation,* takes place as organic matter slowly decays after burial. During the process of decomposition, the organic matter gradually loses its gases and liquids, leaving only a thin film of carbonaceous material. Coal is formed by the same process, and carbonized plant fossils are not uncommon in coal deposits. In addition, some unusual preservations of fish, graptolites and reptiles have been preserved by carbonization.

Petrifaction or Permineralization. Large numbers of fossils have been permineralized or petrified—literally turned to stone. This type of preservation occurs when mineral-bearing ground waters infiltrate porous bone, shell, or plant material. These underground waters deposit their mineral content in the empty spaces of the hard parts, thereby making them denser and more resistant to weathering. Some of the more common minerals deposited in this manner are calcite, silica, and various compounds of iron.

Replacement or Mineralization. Preservation of this type takes place when the original hard parts of organisms are dissolved and removed by underground water. This is accompanied by almost simultaneous deposition of other substances in the resulting voids. Some replaced fossils have had the original structure destroyed by the replacing minerals. Others, as in the case of certain silicified tree trunks, may be preserved in minute detail.

Although more than 50 minerals have been known to replace original organic structures, the most frequent replacing substances are calcite, dolomite (a calcium magnesium carbonate), silica and certain iron compounds.

Traces of Organisms. Fossils consist not only of actual plant and animal remains, but also of any trace or evidence of their existence. Although the latter type of fossil reveals no direct evidence of the original organism, there is some definite indication of the former presence of an ancient plant or animal. This type of fossil may provide much information as to the identity or characteristics of the organism responsible for it.

Moulds and Casts. Shells, bones, leaves and other forms of organic matter are commonly preserved as moulds and casts. If a shell pressed down into the ocean bottom before the sediment hardened into rock, it may have left the impression of the exterior of the shell. This impression is known as a mould. If at some later time this mould was filled with another material, this may have produced a cast. A cast formed in this manner will show the original external characteristics of the shell. Such objects are called external moulds if they show the external features of the hard parts, and internal moulds if the nature of the inner parts is shown.

Moulds and casts are likely to be found in most fossil-bearing rocks. Fossil clams and snails preserved by this method are particularly common even though their shells are composed of minerals that are relatively easy to dissolve, and the original shell material is often destroyed.

Tracks, Trails and Burrows. Many animals have left records of their movements over dry land or the sea bottom. Some of these, for example footprints (Fig. 114), indicate not only the type of animal that left them but may also furnish valuable information about the animal's environment.

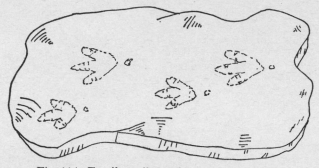

Fig. 114. Fossil reptile tracks preserved in stone

Thus the study of a series of dinosaur tracks may not only indicate the size and shape of the dinosaur's foot but also give some indication as to the weight and length of the animal. Moreover, the type of rock containing the track would probably help determine the conditions under which the animal lived.

A track of half a dozen footprints has been exposed on a wave-cut beach south of Filey in Yorkshire. The size of an individual

Fig. 115. Coprolite

Fig. 116. Gastrolith

footprint is about 15 inches square, comprising three 'toe' impressions and a stump 'heel'. This track is hard to locate, but moulds and other similar footprints are exhibited in Scarborough museum.

Invertebrates also leave tracks and trails of their activities, and markings of this type may be seen on the surfaces of many sandstone and limestone beds. These may be simple tracks, left as the animal moved over the surface, or the burrows of crabs or other burrowing animals. Markings of this sort provide some evidence of the manner of location of these organisms and of the type of environment that they inhabited.

Coprolites. These objects are fossil dung or body waste (Fig. 115). Coprolites may provide valuable information as to the food habits or anatomical structure of the animal that made them.

Gastroliths. These highly polished and well-rounded stones (Fig. 116) are believed to have been used in the stomachs of reptiles for grinding their food into smaller pieces. Large numbers of these 'stomach stones' have been found with the remains of certain types of dinosaurs and other extinct reptiles.

(7) The Classification of Fossils

There are such vast numbers of organisms, both living and extinct, that some system of classification is needed to link them all together. Many fossils bear distinct similarities to plants and animals that are living today, and for this reason palaeontological classification is similar to that used to classify modern organisms. This system, known as the system of **binomial nomenclature,** was proposed in 1758 by Carl Linnæus (or von Linné), a celebrated Swedish naturalist.

Scientific names established in accordance with the principles of binomial nomenclature consist of two parts: the **generic** (or *genus*) name and the **specific,** or **trivial** (*species*) name. These names are commonly derived from Greek or Latin words descriptive of the organism or fossil being named (e.g. *Didymograptus extensus*). They may, however, be derived from the names of people or places, and in such instances the names are always latinized (e.g. *Didymograptus murchisoni*). Greek and Latin are used partly because they are 'dead' languages and not subject to change, but more important, they are also 'international' in that they are recognized by scientists all over the world. The system of binomial nomenclature has led to the development of **taxonomy,** the branch of science dealing with classification.

(8) Units of Classification

The world of organic life has been divided into the plant and animal kingdoms. These kingdoms have been further divided into larger divisions called **phyla** (from the Greek word *phylon*, 'a race'). Each phylum is composed of organisms with certain characteristics in common. For example, all animals with a spinal cord (or notochord) are assigned to the phylum Chordata.

The phylum is reduced to smaller divisions called classes; classes are divided into orders; orders into families; families into genera; and each genus is divided into still smaller units called species. Any species may be further reduced to subspecies, varieties or other subspecific categories.

The generic name and the trivial name together constitute the **scientific name** of a species, and according to this system of classification the scientific name of all living men is *Homo sapiens*. Obviously there are many variations among individual men, but all men have certain general characteristics in common and are therefore placed in the same species.

The following table illustrates the classification of man, a dog and a clam:

	MAN	DOG	CLAM
Kingdom:	Animalia	Animalia	Animalia
Phylum:	Chordata	Chordata	Mollusca
Class:	Mammalia	Mammalia	Pelecypoda
Order:	Primates	Carnivora	Eulamellibranchia
Family:	Hominidae	Canidae	Veneridae
Genus:	*Homo*	*Canis*	*Venus*
Species:	*sapiens*	*familiaris*	*mercenaria*

When writing a scientific name, the generic name should *always* start with a capital (upper-case) letter and the trivial name with a small (lower-case) letter. Both names must be italicized or underlined.

(9) How Fossils are Used

Fossils are useful in a number of ways, for each specimen provides some information about when it lived, where it lived and how it lived.

We use fossils, for example, in tracing the evolution of plants and animals. Fossils in the older rocks are usually primitive and relatively simple, whereas a study of similar specimens that lived in later geologic time reveals that the fossils become progressively complex and more advanced in the younger rocks.

Certain fossils are valuable as **environmental indicators.** For example, the reef-building corals appear to have lived always under much the same conditions as they live today. Hence, if the geologist finds fossil reef corals *in situ* (that is, where they were originally buried), he can be reasonably sure that the rocks containing them were formed from sediments deposited in warm, fairly shallow, salt water. A study of the occurrence and distribution of such marine fossils makes it possible to outline the location and extent of prehistoric seas. The type of fossils present may also provide some indication as to the depth, temperature, bottom conditions and salinity of these ancient bodies of water.

One of the more important uses of fossils is for purposes of **correlation**—the process of demonstrating that certain rock layers are closely related to each other. By correlating or 'matching' the beds containing specific fossils, it is possible to determine the distribution of a geologic formation in a given area. Certain fossils have a very limited vertical or geologic range and a wide horizontal or geographic range. In other words, they lived for only a short time in geologic history but were widely distributed. Fossils of this type are known as **index fossils** or **guide fossils.** They are especially useful in correlation because they are normally associated only with rocks of one particular age.

Microfossils are especially valuable as guide fossils for the petroleum geologists. The **micropalaeontologist** (one who specializes in the study of microfossils) washes the samples brought to the surface by drilling and separates the tiny fossils from the surrounding rock. The specimens are then mounted on special slides and studied under the microscope. Information derived from these minute remains may provide valuable data on the age of the subsurface formation and the possibilities of oil production. In fact, some of the oil-producing zones of Texas and Louisana have even been named after certain key genera of Foraminifera (see page 180). Other microfossils, such as fusulinids, ostracods, spores and pollens, are also used to identify strata and subsurface formations.

Although plant fossils are very useful as climatic indicators, they are not very reliable for purposes of correlation. They do, however, provide much information about the development of plants throughout geologic time.

CHAPTER SIXTEEN

LIFE OF PAST AGES

Some of the plants and animals that inhabited the earth during prehistoric time were surprisingly similar to those living today. But others attained tremendous sizes, assumed bizarre shapes and were quite unlike any modern organism.

The object of this chapter is to supply an introduction to the diversity of life forms of both the past and the present. The taxonomy presented here does not go into great technical detail, but it is adequate for the present purpose, and up-to-date. There are some differences in classification from the schemes given in other, especially older, publications; hence, alternative names are given for a few of the groups listed.

The emphasis here is largely on **morphology,** the study of structure or form. Concentration on this will make it easier to visualize the identifying characteristics of each group studied. It will also help in recognizing some of the more common forms if encountered as fossils. For each group, comments are added on the habitat, most common method of preservation, and geologic range (the known duration of an organism's existence through geologic time).

Appendix C gives, for reference, a simplified classification of plants and animals in outline form.

(1) Plant Classification

Plant fossils are usually fragmental and not well preserved. They have, nevertheless, left a substantial fossil record providing much information about plant evolution. Certain types of plant fossils are also useful as indicators of ancient climatic conditions. Some have an important bearing on the formation of coal.

The following plant classification covers only the larger taxonomic groups. Even so, it may suggest the immense variety of forms which make up the plant kingdom. The term *division* has been used in place of the corresponding term *phylum* applied to the animal kingdom. This usage is now preferred by many botanists and palaeobotanists.

Sub-kingdom Thallophyta. The thallophytes comprise the simplest of all plants. They do not form embryos, nor do they possess roots, stems or leaves. They include such forms as the fungi, algae, lichens and diatoms. The latter are commonly found as microfossils in marine sedimentary rocks (Fig. 117). Some species of algae secrete calcium carbonate in such quantities as to build large limestone masses called reefs. These algal reefs are fairly common in certain

Pre-Cambrian rocks, and are among the oldest fossils known. Thallophytes range from Pre-Cambrian to Recent in age.

Sub-Kingdom Embryophyta. Plants of this sub-kingdom are embryo-forming plants that show considerable advance over the thallophytes. Included here are members of the division **Bryophyta,** the mosses and liverworts. Although they range from Carboniferous to Recent in age, bryophytes are rarely found as fossils.

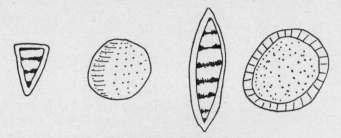

Fig. 117. Diatoms (highly magnified)

Among the more important embryophytes are the plants that have been assigned to the division **Tracheophyta.** The tracheophytes, or vascular plants, have been divided into four subdivisions, among which are many of the more important living and fossil plants. These include such important forms as the ferns, evergreens, flowering plants and hardwood trees. Among the better-known fossil tracheophytes are the cycads, ferns and ginkgos, in addition to such

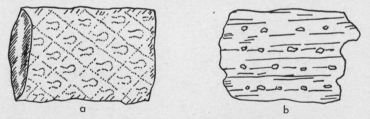

a   b

Fig. 118. Fossil 'coal plants': (*a*) *lepidodendron*, (*b*) *sigillaria*

important coal-forming plants as the club mosses, scouring rushes, and scale trees (Fig. 118). Tracheophytes range from Silurian to Recent in age.

Fossilized seeds, spores, and pollen have also been found. Because of their small size, certain of these are valuable as microfossils.

(2) Animal Classification

The fossilized remains of animals are quite common in many sedimentary rocks. Such remains are of many different kinds, ranging from the shells of microscopic unicellular animals to the bones of huge dinosaurs. The fossils most commonly found, however, are the remains of invertebrate animals such as corals, molluscs and echinoids.

It is not always easy to tell whether certain organisms are plants or animals. Some scientists have suggested that these 'in-betweens' be placed in a separate kingdom called the **Protista.** Members assigned to this kingdom include unicellular organisms such as bac-

Fig. 119. Fossil foraminifers (highly magnified): (*a*) *Fusulina* (carboniferous), (*b*) *Lenticulina* (Cretaceous)

teria, slime moulds, algae, diatoms and protozoans. But only the plant and animal kingdoms will be recognized in this book.

Phylum Protozoa. Members of this phylum are simple unicellular invertebrates (animals without backbones), most of which have no hard parts. Some species, however, have external hard parts that are capable of fossilization. Many protozoans are microscopic in size, and for this reason are important as microfossils.

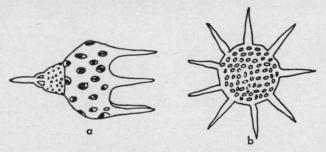

Fig. 120. Fossil radiolarians (highly magnified): (*a*) *Podocyrtis* (Tertiary), (*b*) *Trochodiscus* (Carboniferous)

Included in this phylum is the class Sarcodina, of which two orders, the Foraminifera and Radiolaria, are useful as fossils.

Members of the order Foraminifera (commonly called forams) are predominantly marine animals that secrete tiny many-chambered shells (called tests) of chitin, silica or calcium carbonate (Fig. 119). Forams are abundant in many marine sedimentary rocks and range from Cambrian to Recent in age. Because of their wide distribution and great numbers, the forams are probably the most useful of all microfossils.

The radiolarians secrete delicate, spine-covered, siliceous tests (Fig. 120), and their remains are very abundant in certain oozes or marine sediments. They range from Cambrian to Recent in age, but are not commonly found as fossils.

Phylum Porifera. These are the sponges: the simplest of the many-celled animals. Living sponges secrete chitin, silica, calcium carbonate or spongin in order to form small needle-like hard parts (Fig. 121) called spicules which help to support the soft tissues of the sponge.

Although not abundant as fossils, a few sponges are fairly common in certain Palaeozoic rocks. In addition, the spicules of some species are occasionally found as microfossils. Members of phylum Porifera were probably present during Pre-Cambrian time, and certain sponges and sponge-like forms were abundant during the Cambrian Period.

Fig. 121. Sponge spicules (highly magnified)

Phylum Coelenterata. The coelenterates include a large group of aquatic (water-dwelling) multicelled animals. Although more complex than the sponges, they are rather primitive. The living animal is characterized by a sac-like body cavity, a well-defined mouth, and tentacles which bear stinging capsules. The corals and jellyfishes are typical coelenterates.

Of the three commonly recognized classes of coelenterates—classes Hydrozoa (the hydroids), Scyphozoa (the jellyfishes) and Anthozoa (corals and sea anemones)—only the anthozoans have left a good palaeontological record.

Class Anthozoa. Members of this class are exclusively marine in habitat, and include the sea anemones and corals. The corals are, as we have seen, important geologically. Solitary, or 'horn', corals secrete individual cup- or cone-shaped exoskeletons. Colonial, or compound, corals live together in colonies formed by many individual skeletons attached to each other.

The fossil corals are divided into three groups: Rugosa, Scleractina (Hexacorals) and Tabulate.

The Rugose corals are an extinct group which lived in the Palaeozoic Era. They often formed parts of large reefs and have been used as zone fossils in Carboniferous rocks. Corals, however, are not usually very good zone fossils, as they have a limited distribution. Rugose corals may be solitary (*Zaphrentis*) or colonial (*Lithostrotion*) and exhibit well-defined septa, tabulae and dissepiments (see Fig. 122*a*). Sometimes in the colonial massive forms, the wall between the individual corallities is lost. Common Rugose corals include *Acervularia, Calceola, Caninia, Dibonophyllum, Lithostrotion, Lonsdaleia, Palaeocyclus, Palaeosmilia, Phillipsastraea* and *Zaphrentis*.

The Scleractina, or Hexacorals, are very similar to the Rugose corals in having septa arranged radially. The Scleractina are found,

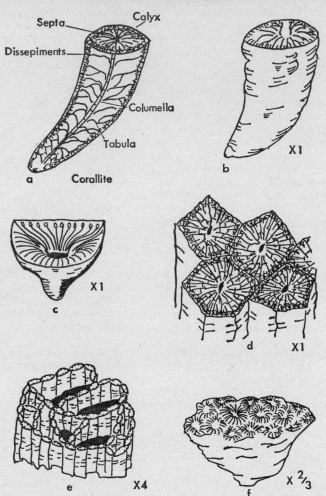

Fig. 122. Corals: (*a*) principal parts, (*b*) *Zaphrentis* (Rugosa—Carboniferous), (*c*) *Calceola* (Rugosa—Devonian), (*d*) *Lithostrotion* (Rugosa—Carboniferous), (*e*) *Halysites* (Tabulate—Silurian), (*f*) *Isastrea* (Scleractina—Jurassic-Cretaceous) (× 1, etc., = magnification).

however, not in Palaeozoic rocks but from the Triassic onwards. It has been suggested, but not proved, that they developed from the Rugose. Common Scleractina include *Goniopora*, *Montlivatia* and *Thamnasteria*.

The Tabulate corals are the oldest known corals, but were overshadowed by the Rugose in the Palaeozoic. They show prominent tabulae, but septa are stunted or absent. All are colonial and may form chains or masses. Common examples include *Alveolites, Favosites, Halysites, Heliolites, Pachypora,Pleurodictyum* and *Syringopora*. They are often included in the order Alcyonaria as they bear a marked resemblance to the modern reef-building corals.

The reef-building corals are, of course, colonial forms and typically live in warm, clear, relatively shallow seas; thus their fossils are good indicators of ancient climates. Although a questionable Cambrian form has been found, corals are definitely known to range from Ordovician to Recent in age and are one of the more important groups of fossils—especially for rocks of Palaeozoic age.

Class Hydrozoa, Order Stromatoporoidea. Although a poor relation to the corals, the Stromatoporoids have at times been important as reef-builders. First found in the Ordovician, they reached their peak in the Devonian, and they have also been found in Mesozoic rocks. They are composed of a calcareous mass of tubes.

Fig. 123. Bryozoans: (*a*) *Archimedes* (Carboniferous), (*b*) *Lioclema* (Carboniferous)

Stromatopora, the commonest member of the order, has been compared with *Millepora*, a prolific reef-builder today and also a Hydrozoan but of a different order.

Worms. As used here, the term 'worms' covers a large group of diverse animals which have been assigned to three phyla: the Platyhelminthes (flatworms), Nemathelminthes (roundworms) and Trochelminthes (rotifers). Because of their lack of hard parts, these animals have left little or no fossil record. This is not the case with the annelid worms, which are discussed on page 191.

Phylum Bryozoa. These small colonial animals, commonly called 'sea mats' or 'moss animals', are abundant in modern seas. Each organism secretes a tiny cuplike exoskeleton of chitinous or calcareous material which may be preserved as a fossil (Fig. 123). Bryozoan fossils are especially abundant in certain Palaeozoic formations. They range from Ordovician to Recent in age, although questionable Cambrian forms have been reported.

Phylum Brachiopoda. The brachiopods (or 'lamp shells') are a large group of exclusively marine animals with shells composed of two pieces called valves (Fig. 124). These valves, typically composed of calcareous or phosphatic material, enclose and protect the soft parts of the animal.

The typical adult brachiopod is attached to the ocean floor by means of a soft fleshy stalk called a pedicle. The pedicle is usually extruded through the pedicle foramen, a hole in the pedicle valve.

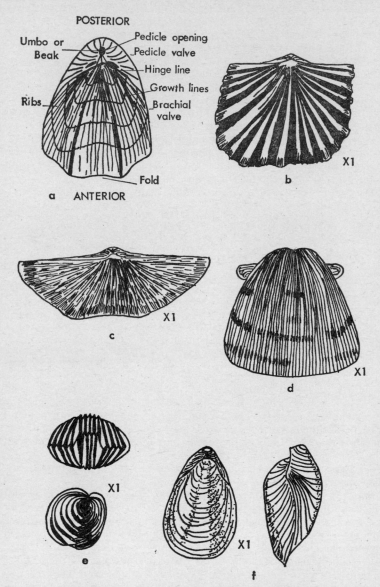

Fig. 124. Brachiopods: (*a*) principal parts of articulate brachiopods (*b*) *Orthis* (Palaeozoic), (*c*) *Spirifer* (Upper Palaeozoic), (*d*) *Productus* (Upper Palaeozoic), (*e*) *Rhynchonella* (Silurian-Recent), (*f*) *Terebratula* (Devonian-Recent)

The other valve of the shell, called the brachial valve, is usually the smaller of the two. (The valves are alternatively referred to as ventral and dorsal, respectively.)

The phylum has been divided into two classes: the Inarticulata and Articulata.

Class Inarticulata. The inarticulate brachiopods are rather primitive forms with a long geologic history stretching from Early Cambrian to Recent. Evidence of their occurrence in Pre-Cambrian rocks has recently been reported. Most are oval to tongue-shaped and lack a pedicle foramen. Their valves are held together by muscles and are not provided with hinge teeth (see below). *Lingula* (Fig. 125) is a typical inarticulate brachiopod. Other common forms are *Crania, Lingulella* and *Orbiculoidea*.

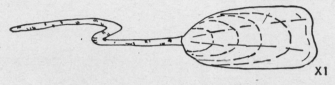

X1

Fig. 125. *Lingula*, an inarticulate brachiopod

Class Articulata. The members of this class have a well-defined hinge, and one valve has well-developed teeth which articulate with sockets in the opposing valve. Articulate brachiopods are characterized by calcareous and phosphatic shells, which are typically of unequal size and assume a wide variety of shapes (Fig. 124). The articulate brachiopods have been divided into two orders and several suborders depending on their distinctive shape and form. Common articulate brachiopods include:

SUBORDER	EXAMPLES
Strophomenacea	*Chonetes, Leptaena, Productus, Strophomena*
Orthocea	*Orthis*
Pentameracea	*Conchidium, Pentamerus, Stricklandia*
Spiriferacea	*Atrypa, Cyrtia, Spirifer*
Rhynchonellacea	*Pugnax, Rhynchonella*
Terebratulacea	*Stringocephalus, Terebratula*

They range from Early Cambrian to Recent in age, and are especially abundant in fossiliferous strata of Palaeozoic age.

Phylum Mollusca. The molluscs are a large group of aquatic and terrestrial (land-dwelling) animals. Members of this phylum include such familiar forms as the snails, slugs, clams, oysters, squids and octopuses. Most molluscs possess a calcareous shell that serves as an exoskeleton, and these hard parts are well adapted for preservation as fossils. But there are some molluscs, for example the slugs, which have no shells, while others (the squids) have internal shells. Because of their relative abundance, great variety, and long geologic history, molluscs are especially common as fossils.

The phylum Mollusca has been divided into five classes: Amphineura (chitons or 'sea-mice'), Scaphopoda ('tusk-shells'), Pelecypoda (clams and oysters), Gastropoda (snails and slugs), and Cephalopoda (squids, octopuses and the extinct ammonoids). Only the classes

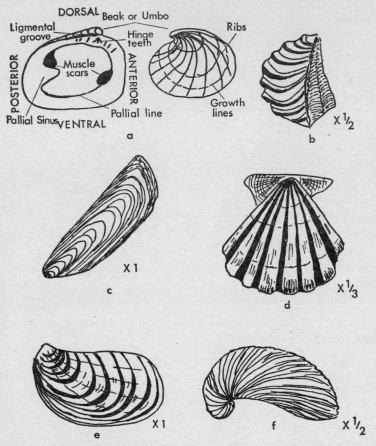

Fig. 126. Lamellibranchs (pelecypods): (*a*) principal parts, (*b*) *Trigonia* (Jurassic-Recent), (*c*) *Modiola* (Devonian-Recent), (*d*) *Pecten* (Triassic-Recent), (*e*) *Inoceramus* (Jurassic-Cretaceous), (*f*) *Gryphaea* (Jurassic-Eocene)

Pelecypoda, Gastropoda and Cephalopoda are commonly found as fossils.

Class Pelecypoda. The lamellibranchs (or pelecypods as they are known in America) are characterized by a shell composed of two calcareous valves (Fig. 126a) which enclose the soft parts of the animal. They are exclusively aquatic and may be found in both fresh and salt water. Most lamellibranchs are slow-moving bottom-dwelling forms like the clams, but some, for example the oysters, are attached. Others, like the scallops, are swimmers.

The typical lamellibranch shell is composed of two valves of equal size and form held together by a tough elastic ligament which runs along the dorsal (top) side of the shell. In addition, most forms have teeth and sockets which are located along the hinge line. The exterior of the shell is commonly covered by a horny outer covering called the periostracum. The inner surface of each valve is lined with a calcareous layer of porcellaneous or pearly material.

Although the shell form may vary considerably (Fig. 126), most lamellibranchs are clamlike. The beak, which represents the oldest part of the shell, is located on the anterior (front) end of the shell (the rear end of the shell is designated the posterior). When the shell is elongated posteriorly, it is an indication that the lamellibranch was a partially buried form. The lower margin of the shell (where the valves open) is called the ventral margin. As noted earlier, the dorsal margin is the top side along which the hinge and ligament are located.

The interior of the valve may be marked by such structures as teeth, sockets, muscle scars, and other diagnostic features. The exterior of many shells is marked by concentric growth lines, as well as nodes, spines, ribs, and other types of ornamentation.

Fossil lamellibranchs are commonly preserved as casts and moulds, but many are found with original shell material that appears to have undergone little change.

The first lamellibranchs appeared in the Ordovician Period and were especially abundant during Late Palaeozoic, Mesozoic, and Cainozoic times.

Like the brachiopods, the lamellibranchs are numerous and sub-divided into various orders and suborders, depending on various anatomical details such as the hinge line. The forms of lamellibranchs are legion, but a few of the commoner ones include *Astarte, Carbonicola, Cardiola, Cardita, Cardium, Exogyra, Glycimeris, Gryphaea, Inoceramus, Lima, Mya, Ostrea, Pecten, Spondylus, Trigonia* and *Venus*.

Class Gastropoda. The typical gastropod has a spirally coiled, single-valved, unchambered shell. Most gastropods have gills and live in shallow marine waters, but some inhabit fresh water. Others are land-dwellers and breathe by means of lungs.

Gastropods range from Early Cambrian to Recent, and both fossil and modern gastropods exhibit a variety of shapes, sizes and ornamentation (Fig. 127). Such shells may be flat, spirally coiled, cone-shaped, turreted or cylindrical. The closed, pointed end of the shell is known as the apex, and each turn of the shell is called a whorl. The largest and last-formed Whorl is called the body whorl, and the opening of the shell (called the aperture) is in this whorl. The 'spire' is composed of the combined whorls exclusive of the body whorl. In the majority of cases the coiling is *dextral* (to the right), but a few *sinistral* forms do occur either as 'sports' or in a particular species or genus. The inner and outer margins of the aperture are designated the inner lip and the outer lip, respectively. Some snails close the aperture by means of a horny or calcareous plate, called the operculum, attached to the foot of the animal. This plate effectively seals the aperture when the animal is withdrawn into its shell.

Large numbers of gastropods, especially certain Mesozoic and Palaeozoic forms, have been preserved as internal or external moulds.

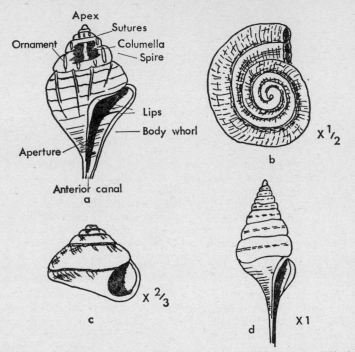

Fig. 127. Gastropods: (*a*) principal parts, (*b*) *Euomphalus* (Silurian-Triassic), (*c*) *Pleurotomaria* (Triassic-Recent), (*d*) *Fusinus* (Cretaceous-Recent)

Internal moulds are formed after the animal dies and its soft parts have decomposed. This enables the shell to become filled with sediment, which later becomes solidified. The shell may eventually be removed by weathering or solution, freeing the internal mould. This type of mould is called a *steinkern* and normally does not reveal any external shell characteristics. Common and distinctive gastropods include *Bellerophon, Conorbis, Euomphalus, Fusinus, Planorbis, Turritella, Viviporus* and *Voluta*.

Class Cephalopoda. The cephalopods are marine molluscs with or without shells. When a shell is present, it may be internal or external, chambered or solid. They are exclusively marine, carnivorous (meat-eating), unattached animals with a high degree of body organization. Cephalopods are among the most advanced of all molluscs and include the squid, octopus, pearly nautilus, and the extinct ammonoids. Their geologic range is from Cambrian to Recent, but they were much more abundant in ancient seas than they are today.

The class is usually divided into three subclasses: the Nautiloidea (represented today by the pearly nautilus), Ammonoidea (the extinct ammonoids) and the Coleoidea (octopuses and squids). Of these, the nautiloids and ammonoids constitute important fossil groups.

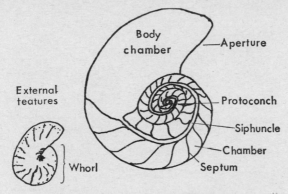

Fig. 128. Morphology and principal parts of pearly nautilus

Subclass Nautiloidea. The nautiloids are cephalopods with external chambered shells in which the septa (dividing partitions) are simple and have smooth edges. This subclass is represented by a single living form, *Nautilus*, and a large number of fossil species.

Nautilus has a shell composed of calcium carbonate, which is coiled in a flat spiral (Fig. 128). The interior of the shell is divided into a series of chambers by calcareous partitions called septa. The juncture of the septum and the inner surface of the shell forms what is known as the suture. These suture lines (Fig. 129) cannot be seen unless the outer shell has been removed, but they are visible on the internal moulds of many fossil cephalopods, and are important in nautiloid and ammonoid classification. Nautiloids are characterized

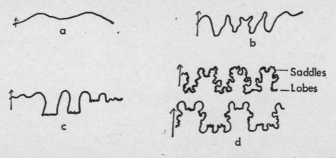

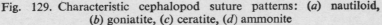

Fig. 129. Characteristic cephalopod suture patterns: (*a*) nautiloid, (*b*) goniatite, (*c*) ceratite, (*d*) ammonite

by simple, smoothly curved suture patterns, but ammonoids display more complex and wrinkled sutures.

Although the shell of the only living nautiloid is coiled, many of the earlier forms had uncoiled, cone-shaped shells. Some of the early Palaeozoic forms measured as much as 15 feet long.

The earliest known nautiloids have been collected from Lower

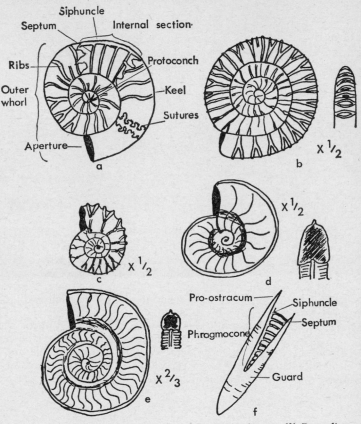

Fig. 130. Ammonites and a belemnite: (*a*) principal parts, (*b*) *Dactylioceras* (Jurassic), (*c*) *Schloenbachia* (Cretaceous), (*d*) *Harpoceras* (Jurassic), (*e*) *Hildoceras* (Jurassic), (*f*) a belemnite

Cambrian rocks, and they were much more abundant during the Palaeozoic Era than they are today. Orthoceras was a common straight or slightly curved form which existed from the Ordovician to the Triassic.

Subclass Ammonoidea. Members of this subclass are extinct cephalopods which are related to the nautiloids but characterized by more complex suture patterns. Ammonoids have an external partitioned shell which may be straight, curved, or spirally coiled (Fig. 130). They appeared first in Devonian times, were abundant and varied during the Mesozoic, and were extinct by the end of the Cretaceous Period.

Ammonoids exhibit three types of sutures: (*a*) **goniatitic** (a curved and angular pattern), (*b*) **ceratitic** (curved and crenellated—marked in places by a series of toothlike indentations), and (*c*) **ammonitic** (a very complex subdivided pattern). Cephalopods with ammonitic sutures range from Carboniferous to Cretaceous in age and were the most abundant cephalopods of the Mesozoic. The Clymeniids, characterized by *Clymenia*, are a unique group which coiled in the opposite sense to all other ammonites, and have the siphuncle (see Fig. 128) at the internal margin.

The Goniatitids characterized by *Goniatites* have a simple suture line.

The few ammonites mentioned below are just some of the common forms: *Amaltheus, Ceratities, Dactylioceras, Hamites, Harpoceras, Hildoceras, Hoplites, Lytoceras, Phylloceras, Schloenbachia, Stephanoceras* and *Turrilites* (a spiral form).

Subclass Coleoidea. These are cephalopods that are characterized by an internal shell or no shell at all. Included here are the squids, octopuses and cuttlefish, and the extinct belemnites. The belemnites (also called belemnoids) appear to represent the fossil remains of a living animal that was similar to the modern cuttlefish. Their characteristic shape has caused them to be called 'finger stones' or 'fossil cigars'. They range from Carboniferous to Cretaceous in age, and certain species are valuable guide fossils for Jurassic and Cretaceous rocks.

Phylum Annelida. The annelid worms are segmented worms such as the common earthworm. They are marine, fresh-water or terrestrial in habitat, and have apparently been common throughout much of geologic time. Because of their lack of hard parts, annelids have left little direct evidence of their activities in the geologic past. Some, however, have left calcareous tubes, chitinous jaws and teeth called scolecodonts, and burrows and borings.

Undoubted fossil annelid remains have been found in rocks ranging from Cambrian to Recent in age. However, the presence of wormlike tracks and burrows in certain Pre-Cambrian rocks suggests that annelids probably existed before the Cambrian Period.

Phylum Arthropoda. The arthropods are one of the more advanced groups of invertebrates, and they are known from the Cambrian to the Recent. Modern representatives of this group include the crabs, shrimps, crayfish, insects, and spiders. Arthropods vary greatly in size and shape and are among the most abundant of all animals.

They have adapted themselves to a variety of habitats and live on land, in the water, and in the air. Although of great importance in nature today, only a few groups are of interest to the palaeontologist. Three of these are the trilobites, ostracods, and eurypterids.

Trilobites. These are members of the class Trilobita of the subphylum Trilobitomorpha. The trilobites are exclusively marine arthropods, which derive their name from the typical three-lobed appearance of their bodies. The body of the animal is encased in a chitinous exoskeleton and is divided into a central or axial lobe, and two lateral or pleural lobes (Fig. 131). The body is also divided, from front to back, into the cephalon (head), the thorax (abdomen), and the pygidium (tail). Some species have the body segments of the thorax arranged in such a manner as to permit the trilobite to roll up into a ball, and many trilobites are found in this position (Fig. 131). Various attempts have been made to subdivide the trilobites

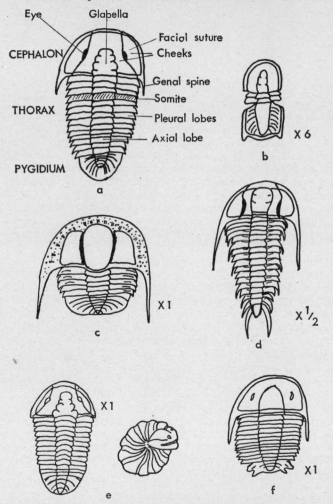

Fig. 131. Trilobites: (*a*) principal parts, (*b*) *Agnostus* (Lower Palaeozoic), (*c*) *Trinucleus* (Ordovician), (*d*) *Paradoxides* (Cambrian), (*e*) *Calymene* (Silurian), (*f*) *Angelina* (Ordovician)

according to whether they are isopygous (with cranidium and pygidium the same size); or whether they are blind (have no eyes); or by the position of the facial suture. A few of the common trilobites are *Acidaspis*, *Agnostus*, *Angelina*, *Asaphus*, *Calymene*, *Olenellus*, *Olenus*, *Paradoxides*, *Phacops* and *Trinucleus*. Trilobites range from Cambrian to Permian in age, and were especially abundant during some of the earlier Palaeozoic periods.

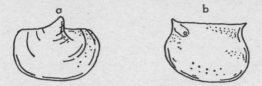

Fig. 132. Palaeozoic ostracods (highly magnified): (*a*) *Paraechmina*, (*b*) *Leperditia*

Ostracods. These minute bivalved arthropods belong to the class Ostracoda of the sub-phylum Crustacea—the same sub-phylum to which the crabs, shrimp, crayfish, and lobsters have been referred. The ostracods are small aquatic crustaceans which externally resemble small clams (Fig. 132). However, the animal that inhabits the shell possesses all of the typical arthropod characteristics. Fossil ostracods range from Ordovician to Recent in age, and because of their small size they are especially useful as microfossils.

Eurypterids. These extinct arthropods have been assigned to the class Merostomata of the sub-phylum Chelicerata. The scorpions,

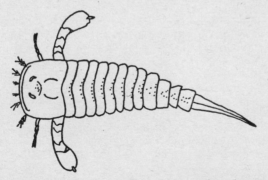

Fig. 133. Eurypterid, an Early Palaeozoic arthropod

spiders, mites, ticks and 'king crabs' have also been assigned to this sub-phylum. Eurypterids were scorpion-like aquatic forms characterized by broad winglike appendages (Fig. 133). Some are believed to have possessed a poison gland and sting.

Although not common as fossils, they occur locally in certain Silurian and Devonian formations. Their geologic range is from Early Ordovician to Permian.

Phylum Echinodermata. The echinoderms are a large group of exclusively marine animals, most of which exhibit a marked fivefold symmetry. The typical echinoderm is a relatively complex organism with a skeleton composed of numerous calcareous plates which are intricately fitted together and covered by a leathery outer skin called the integument. Many echinoderms have a typical star-shaped body, but some types may be heart-shaped, bun-shaped, or cucumber-shaped. Echinoderms are abundant as fossils in marine sedimentary rocks of all ages.

The phylum has been divided into two sub-phyla: the Pelmatozoa (attached echinoderms), and Eleutherozoa (unattached echinoderms).

Sub-phylum Pelmatozoa. These are echinoderms which are more or less permanently attached to the sea bottom by means of a stalk

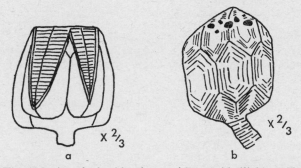

Fig. 134. Attached echinoderms: (*a*) cystoid, (*b*) blastoid

composed of slightly movable, calcareous, disc-like segments. They range from Cambrian to Recent in age and are particularly common as fossils in Palaeozoic rocks, although the only case where they are used as a zone fossil is in the Chalk, which is Mesozoic.

Although the Pelmatozoa has been divided into several classes, only three of these, the Cystoidea, Blastoidea and Crinoidea need be discussed here. With the exception of the Crinoidea, all of the pelmatozoans are extinct.

Class Cystoidea. These are primitive, extinct, attached echinoderms which were relatively common during the early Palaeozoic. Cystoids are characterized by a globular or sac-like calyx (the main body skeleton), which is made up of numerous irregularly arranged calcareous plates (Fig. 134*a*). The animal was attached to the sea floor by a short stem.

Ranging from Cambrian to Devonian in age, cystoids were especially numerous during the Ordovician and Silurian Periods.

Class Blastoidea. Members of this extinct class are short-stemmed echinoderms with a small, symmetrical bud-like calyx. The mouth is located near the centre of the calyx and is surrounded by five openings called spiracles. Five distinct 'food grooves' radiate outward from the mouth.

Blastoids range from Ordovician to Permian in age and were especially abundant during the Carboniferous Period.

Class Crinoidea. The crinoids, commonly called sea-lilies because of their flower-like appearance, are the only attached echinoderms living today. The crinoid calyx is composed of numerous symmetrically arranged plates, and most crinoids have a long stem

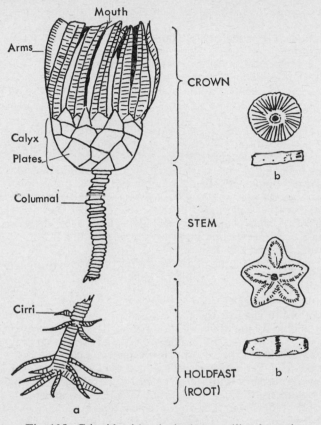

Fig. 135. Crinoids: (*a*) principal parts, (*b*) columnals

(Fig. 135). Other crinoids are free-swimming in the adult stage and are attached only during the earlier phases of their development.

The typical crinoid has a cup-shaped calyx with five grooves radiating outward from its centre. These grooves, used as channels to convey food to the mouth, continue outward along the complexly segmented arms. The stem consists of a relatively long flexible stalk composed of numerous calcareous disc-shaped segments called columnals. Each columnal contains a round or star-shaped opening in its centre. These columnals become separated when the animal dies, and certain Palaeozoic limestones contain such great numbers of columnals that they are called crinoidal limestones.

The crinoid stem is attached to the sea floor by means of the 'hold-fast', a structure which branches out into the surrounding sediments and serves to anchor the animal to the bottom.

Crinoids range from Ordovician to Recent in age, and their remains are particularly abundant in Palaeozoic strata. The majority of the crinoids living today are stemless free-swimming forms called 'feather-stars'.

Sub-phylum Eleutherozoa. The eleutherozoans are unattached, free-swimming bottom-dwelling echinoderms. The sub-phylum has been divided into three classes: the Stelleroidea (starfishes and 'brittle stars'), Echinoidea (sea-urchins), and Holothuroidea (sea-cucumbers). Of these, only the Echinoidea are useful as fossils.

Class Stelleroidea. The stelleroids are typically star-shaped, free-moving echinoderms, and include such familiar forms as the starfishes and 'serpent stars' or 'brittle stars'. Although none of this group is a particularly useful fossil, they do have a long geologic history from Ordovician to Recent.

Class Echinoidea. The echinoids are a group of unattached echinoderms with bodies consisting of numerous calcareous plates and spines. They do not possess the radiating armlike extensions which characterize the stelleroids; instead, their bodies are disc-shaped, heart-shaped, bun-shaped or globular (Fig. 136). Modern representatives of the Echinoidea include such familiar forms as the heart-urchins, and sea-urchins.

The test (exoskeleton) of the echinoid is composed of many intricately fitting calcareous plates which enclose and protect the animal's soft parts. The outside of the test is commonly covered with a large number of spines of varying sizes which are used to aid locomotion, to support the test, and for purposes of protection.

The echinoids are divided into two orders: Regularia and Irregularia. The Regular echinoids are circular, with the anus in the centre, and show a distinct radial symmetry. They are found from the Ordovician Period onwards, and include *Archaeocidaris*, *Cidaris*, *Echinus*, *Hemicidaris* and *Melonechinus*. The Irregular echinoids lack radial symmetry (owing to migration of the anus, and often movement of the mouth) but they do show a bilateral symmetry. Not found in the Palaeozoic, the Irregularia become very abundant in the Jurassic and Cretaceous, and are used extensively as zone fossils in the chalk. Eight of the commonest are *Clypeaster*, *Conulus*, *Discoidea*, *Echinocorys*, *Holaster*, *Holectypus*, *Micraster* and *Pygaster*.

Class Holothuroidea. The sea-cucumbers are characterized by a sac-like, cucumber-shaped body. With the exception of small calcareous rods or plates called sclerites or ossicles, holothuroids do not possess any hard parts and are therefore seldom fossilized. Sclerites as old as Carboniferous have been reported, and impressions of forms resembling sea-cucumbers have been described from Middle Cambrian rocks in Canada. The sea-cucumbers make up only a small fraction of our present-day marine fauna.

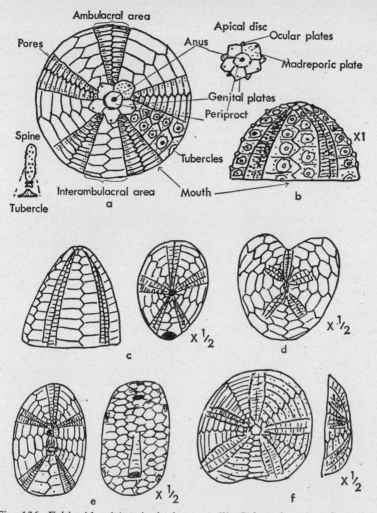

Fig. 136. Echinoids: (*a*) principal parts, (*b*) *Cidaris* (regular—Jurassic-Cretaceous), (*c*) *Conulus* (irregular—Cretaceous), (*d*) *Micraster* (irregular—Cretaceous), (*e*) *Echinocorys* (irregular—Cretaceous) (*f*) *Clypeus* (irregular—Jurassic)

Phylum Chordata. Members of this phylum are characterized by the presence of a well-developed nervous system and a body supported by a bony or cartilaginous notochord or spinal column.

Only two of the several chordate sub-phyla are of palaeontological significance: the Hemichordata (which includes the extinct graptolites) and the Vertebrata (including all animals with backbones).

Sub-phylum Hemichordata. Hemichordates do not possess a true backbone; rather, they are characterized by a well-defined notochord which runs the length of the body. Only the class Graptolithina is of palaeontological importance.

Class Graptolithina. The graptolites are a group of extinct colonial animals that were very abundant during Early Palaeozoic times.

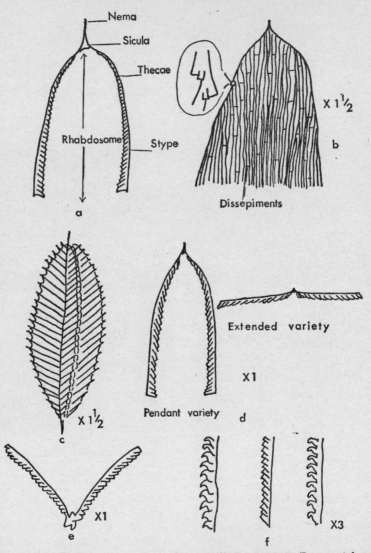

Fig. 137. Graptolites: (a) principal parts, (b) *Dictyonema* (Lower Palaeozoic), (c) *Phyllograptus* (Ordovician), (d) *Didymograptus* (Ordovician), (e) *Dicellograptus* (Ordovician), (f) *Monograptus* (Silurian)

They are characterized by a chitinous exoskeleton, consisting of rows of cups or tubes, which housed the living animal. These grew along single or branching stalks (Fig. 137) which were attached to rocks, seaweeds, or other foreign objects. Some graptolites were attached to floating objects and attained wide distribution in this manner.

The Graptolithina are subdivided into several orders, of which Dendroidea and Graptolitoidea are the most important. The Dendroidea are multibranched graptolites with three distinct types of *thecae* (sheaths). Sometimes the branches are interconnected with dissepiments. Ranging from Cambrian to Devonian in age, the commonest examples are *Callograptus, Clonograptus, Dendrograptus* and *Dictyonema*. The Graptolitoidea have fewer branches (usually four, two or one), and far fewer thecae. They are largely confined to Ordovician and Silurian rocks. These graptolites include many common forms, notably *Bryograptus, Climacograptus, Cyrtograptus, Dicellograptus, Dicranograptus, Didymograptus, Diplograptus, Monograptus, Phyllograptus, Rastrites* and *Retiolites*.

There has been considerable uncertainty about the exact taxonomic position of these animals, and older classifications have considered them to be members of the classes Hydrozoa, Scyphozoa and Graptozoa of phylum Coelenterata. Some early palaeontologists placed them in the phylum Bryozoa. Recent research, however, suggests that the graptolites are some type of extinct chordate.

Graptolites are recorded in rocks ranging from Cambrian to Carboniferous in age. They are especially useful as guide fossils in certain black shales of the Ordovician and Silurian ages, where their remains are commonly preserved as a flattened carbon residue.

Sub-phylum Vertebrata. The vertebrates, the most advanced of all chordates, are characterized by a skull, a bony or cartilaginous internal skeleton, and a vertebral column of bone or cartilage.

The sub-phylum is commonly divided into two superclasses, the Pisces (fishes and their relatives) and the Tetrapoda (the 'four footed' animals—including snakes, which have vestigial limbs).

Although the geologist makes most of his palaeontological correlations with invertebrate fossils, the remains of vertebrates have provided much needed and valuable palaeontological information. Furthermore, such fossils as the remains of dinosaurs, giant fishes, mammoths, sabre-toothed tigers and other highly distinctive vertebrates, are among the most interesting and fascinating known.

Superclass Pisces. The Pisces, commonly called fishes, are the simplest of the vertebrate animals. They are aquatic, free-moving, and cold-blooded—meaning that their blood is maintained at the temperature of the surrounding water. Most forms breathe by means of gills. (The unusual lungfishes of Australia and Africa are exceptions: they breathe by means of a primitive lung developed from the swim bladder.)

The four classes recognized in most recent fish classifications are briefly considered below.

Class Agnatha. These are primitive jawless fishes, represented by such living forms as the lampreys and hagfishes. The earliest agnathans, the ostracoderms, first appeared in the Ordovician and were extinct by the end of the Devonian. These early fishes, which were armoured by a bony covering on the front part of their bodies, are believed to be the earliest known vertebrates.

Class Placodermi. Placoderms are primitive jawed fishes, most of which were heavily armoured. These creatures were sharklike in appearance, and some grew to be as much as 30 feet in length. They first appeared in the Silurian and were extinct by the end of the Permian.

Class Chondrichthyes. Members of this class are characterized by a skeleton composed of cartilage. Included here are the sharks, skates,

Fig. 138. Fossil shark teeth

and rays. They first appeared in Devonian times and have been relatively common up to the present. Fossil shark teeth (Fig. 138) are relatively abundant in certain Mesozoic and Cainozoic formations; for example, in Britain they are found in the Eocene Barton Clay of Hampshire.

Class Osteichthyes. These are the true bony fishes—the most highly developed and abundant of all fishes. They are typified by well-developed jaws, an internal bony skeleton, an air bladder, and,

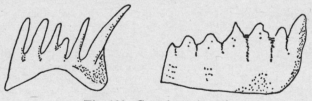

Fig. 139. Conodonts (× 30)

typically, an external covering of overlapping scales. Fish fossils are commonly found in the form of teeth, bones and scales, with occasionally a well preserved skeleton. They range from Middle Devonian to Recent in age.

The small, amber-coloured, toothlike fossils called **conodonts** (Fig. 139) are believed to be the hard parts of some type of extinct fish. Although palaeontologists are not certain what type of animal these strange fossils represent, their small size and rather limited stratigraphic occurrence makes them useful as microfossils.

In the Devonian a group of lobe-finned crossopterygians became prominent, and among them were the coelacanths. The group was thought to have become extinct about 70 million years ago, but in 1938 a coelacanth was caught in the Mozambique Channel off South Africa. Since then a few other specimens have been found and preserved.

Superclass Tetrapoda. These are the most advanced chordates, and characteristically possess lungs, a three- or four-chambered heart, and paired appendages. Four classes are recognized: the Amphibia (frogs, toads and salamanders), Reptilia (lizards, snakes and turtles), Aves (birds), and Mammalia (mammals, such as men, dogs, bats, whales).

Class Amphibia. The amphibians, represented by such common forms as the toads, frogs and salamanders, are cold-blooded animals

Fig. 140. Amphibians: (*a*) *Eryops* (Permian), (*b*) *Dimetrodon* (Permian)

that breathe primarily by lungs. They spend most of their life on land, but during their early stages of development they live in the water, where they breathe by means of gills.

The amphibians were the earliest developed four-legged animals and appeared late in the Devonian Period. Their fossils are relatively abundant in the Carboniferous, Permian and Triassic rocks (Fig. 140). The amphibians made a great step forward by invading the land, but they had to return to the water to lay their eggs. Later, some forms returned to a purely aquatic life. The ability to lay eggs with a shell, which necessitated internal fertilization, was achieved by their descendants the reptiles. Some forms possess both amphibian and reptilian characteristics, and are therefore difficult to assign to either group unambiguously.

Class Reptilia. The reptiles evolved from the amphibians. They have become adapted to permanent life on land, and need not rely on an aquatic environment. They are cold-blooded and are normally characterized by a scaly skin. Reptiles were formerly much more abundant than they are today, and they assumed many different shapes and sizes in the geologic past. Modern classifications recognize a large number of reptilian groups, but only the more important ones are briefly reviewed here.

Cotylosaurs. These primitive reptiles, although retaining some amphibian characteristics, became adapted to an exclusively terrestrial mode of life. They lived during the Carboniferous and Permian Periods, and apparently became extinct during Triassic times.

Turtles and Tortoises. The bodies of these reptiles are more or less completely encased by bony plates. They range from Late Triassic to Recent in age, although there is questionable evidence of a Late Permian turtle-like reptile. Some of the Cainozoic tortoises were three to four feet long, and certain of the Cretaceous sea turtles attained lengths of as much as 12 feet.

Pelycosaurs. These are a group of Late Palaeozoic reptiles, some of which were characterized by the presence of a sail-like fin on their backs.

Therapsids. The therapsids constitute a mammal-like group of reptiles which were well adapted to a terrestrial existence. Although this group is not especially important as fossils, study of their remains has provided much information about the origin of the mammals. They range from Middle Permian to Middle Jurassic in age.

Ichthyosaurs. The so-called 'fish-lizards' were short-necked marine reptiles which were definitely fishlike in appearance. They resembled the modern dolphins and some of them attained lengths of 25–30 feet, though the average was much less. Ichthyosaurs first appeared in Middle Triassic times and were extinct by the end of the Cretaceous.

Mosasaurs. The mosasaurs, another group of extinct marine lizards, grew to be as much as 50 feet long. That they were vicious carnivores is indicated by their great, gaping, tooth-filled jaws. These giant reptiles lived only during the Cretaceous Period.

Plesiosaurs. The plesiosaurs were marine reptiles which were characterized by a broad turtle-like body, paddle-like flippers, and a long neck and tail. Although these reptiles were not as streamlined or well equipped for swimming as the ichthyosaurs or mosasaurs, their long serpent-like necks were probably quite useful in catching fish and other small animals for food. Plesiosaur remains range from Middle Triassic to Late Cretaceous in age.

Phytosaurs. The phytosaurs were a group of crocodile-like reptiles, which ranged from 6 to 20 feet in length. Although they resembled crocodiles both in appearance and in their mode of life, this similarity is only superficial, for the phytosaurs and crocodiles are two distinct groups of reptiles. The phytosaurs lived only during the Triassic Period.

Crocodiles and Alligators. These reptiles, which are common today, adapted themselves to the same type of habitat as that occupied by the phytosaurs, which preceded them. Crocodiles and alligators were much larger and more abundant during Cretaceous and Cainozoic times than they are today. The crocodiles first appeared in the Cretaceous and the alligators in the Tertiary.

Pterosaurs. These were Mesozoic reptiles with batlike wings supported by arms and long thin 'fingers' (Fig. 141). The pterosaurs were

Fig. 141. Pterosaur, a flying reptile: *Rhamphorynchus* (Jurassic)

well adapted to life in the air, and their lightweight bodies and wide skin-covered wings enabled them to soar or glide for great distances. The earliest known pterosaurs were collected from Lower Jurassic rocks, and the group had become extinct by the end of the Cretaceous Period. During the Cretaceous, certain of these creatures attained a wingspread of as much as 27 feet, but their bodies were small and light.

Dinosaurs. The collective term 'dinosaurs' (which literally means 'terrible lizards') has been given to a group of reptiles that dominated the life of the Mesozoic for some 165 million years. These strange creatures ranged from as little as a few feet to as much as 85 feet in length and from a few pounds to perhaps 45 tons in weight. Some dinosaurs were carnivorous, but most were herbivorous (plant-eaters). Some forms were bipedal (walked on their hind legs), while others were quadrupedal (walked on all fours), and although most of the dinosaurs were land-dwelling forms, aquatic and semi-aquatic species were also present.

The dinosaurs have been divided into two orders based on the structure of their hip-bones: the Saurischia (those with a lizard-like pelvic girdle) and the Ornithischia (those with a birdlike pelvic girdle).

Saurischian Dinosaurs were particularly abundant during the Jurassic, and are characterized by hip-bones that are similar to those of the modern lizards. Saurischian dinosaurs were first discovered in Triassic rocks, and they did not become extinct until the end of Cretaceous time. The lizard-hipped reptiles of order Saurischia have been divided into two rather specialized suborders: the Theropoda (carnivorous bipedal dinosaurs that varied greatly in size) and the

Fig. 142. *Tyrannosaurus rex*, a Cretaceous carnivore

Sauropoda (herbivorous, quadrupedal, semi-aquatic, usually gigantic dinosaurs).

Theropods. These saurischian dinosaurs walked on birdlike hind limbs, and they were exclusively meat-eating forms. Some of the theropods were exceptionally large and were undoubtedly vicious beasts of prey. This assertion is borne out by such characteristics as the small front limbs with long sharp claws for holding and tearing flesh, and the large strong jaws which were armed with numerous sharp teeth. The largest of all known theropods was *Tyrannosaurus rex* which, when standing on its hind limbs, was almost 20 feet tall (Fig. 142). Some were as much as 50 feet long. *Tyrannosaurus* is believed to have been among the most savage animals ever to inhabit the earth.

Sauropods. The sauropods included the largest dinosaurs. Some, such as *Brontosaurus* (Fig. 143) were 85 feet long and probably weighed 40–50 tons. The sauropods were primarily herbivorous dinosaurs which had become adapted to an aquatic or semi-aquatic way of life, and they probably inhabited lakes, rivers and swamps.

Ornithischian Dinosaurs, or bird-hipped dinosaurs, were herbivorous reptiles which varied widely in form and size. They appear to have been more highly developed than the saurischians. Included in this order are the duck-billed dinosaurs (suborder Ornithopoda), the

Fig. 143. *Brontosaurus*, a Mesozoic dinosaur

plate-bearing dinosaurs (suborder Stegosauria), the armoured dinosaurs (suborder Ankylosauria), and the horned dinosaurs (suborder Ceratopsia).

Ornithopods. The ornithopods were unusual dinosaurs which were predominantly bipedal and semi-aquatic. Some (like *Trachodon*, the duck-billed dinosaur) were highly specialized.

Stegosaurs. These were herbivorous, quadrupedal ornithischians with large projecting plates along their backs and heavy spikes on their tails. *Stegosaurus*, a Jurassic form, is typical of the plate-bearing dinosaurs (Fig. 144). This great beast weighed about 10 tons, was some 30 feet long, and stood approximately 10 feet tall at the hips. *Stegosaurus* was characterized by a double row of large, heavy, pointed plates running along its back. These plates began at the back of the skull and stopped near the end of the tail. The tail was also equipped with four or more long curved spikes which were probably used for purposes of defence. The animal had a very small skull housing a brain about the size of a walnut. It is assumed that these, and all other dinosaurs, were of very limited intelligence.

Ankylosaurs. The ankylosaurs were four-footed, plant-eating Cretaceous reptiles characterized by relatively flat bodies. The skull and back of the dinosaur were protected by bony armour, and the clublike tail was armed with spikes. *Ankylosaurus*, a typical ankylosaur, also had spines projecting from along the sides of the body and tail. The armoured spiked back and the heavy clublike tail probably provided the animal with much-needed protection against the predatory carnivorous dinosaurs.

Fig. 144. *Stegosaurus*, plate-back dinosaur

Ceratopsians. The horned dinosaurs, or ceratopsians, are another group of dinosaurs that are exclusively Cretaceous in age. These herbivorous creatures had beaklike jaws, a bony neck frill which extended back from the skull, and one or more horns. *Triceratops* was the largest of the horned dinosaurs. Some forms were 30 feet long, and the skull could measure as much as 8 feet from the tip of the parrot-like beak to the back of the neck shield.

Class Aves. Birds, because of the fragile nature of their bodies, are seldom found as fossils. Nevertheless, some interesting and important fossil birds have been discovered. The earliest of them was found in rocks of Late Jurassic age in Germany. Named *Archaeopteryx*, this primitive bird was little more than a reptile with feathers, and was about the size of a pigeon. It possessed a lizard-like tail, toothed beak, and certain other definitely reptilian characteristics.

Birds underwent numerous changes during Cretaceous times, and most of our present-day forms had developed by the end of the Tertiary.

Class Mammalia. The mammals are the most advanced of all vertebrates. They are born alive and fed with milk from the mother's breast, are warm-blooded, air-breathing creatures characterized by a protective covering of hair. (These characteristics are the most typical mammalian features, but exceptions are found in certain mammals.)

First appearing in Jurassic times, mammals were probably derived from some form of mammal-like reptile. They were rare during the Mesozoic, but underwent rapid development and expansion in numbers during the Cainozoic, when certain types of mammals become extremely large and assumed many bizarre shapes. The majority of these were doomed to early extinction but are well known from their fossils. Two of the primitive forms of mammals, the monotremes and marsupials, are still represented by living forms.

Monotremes probably evolved in the Jurassic. They exhibited various reptilian characteristics, being, for instance, the only mammals that ever laid eggs. They fed their young on milk secreted from a modified sweat gland, the forerunner of the female mammal's breasts. Today the monotremes are confined to Australia, where they are represented by the duck-billed platypus and spiny anteaters.

Marsupials first appeared in the Cretaceous Period, when they were rather like opossums. The main characteristic of the marsupials is that they produce live young which are very poorly developed, and complete their 'babyhood' attached to milk teats in their mother's pouch. Tiger-like sabre-toothed varieties occurred in the Cainozoic, but gradually their numbers declined and forms became fewer. Outside Australia, the only remaining marsupials are the opossums of the Americas. The most famous of the Australian marsupials is the kangaroo. Other well known forms are the koala bear, wallaby and wombat. Among the less well known are the bandicoot and the Tasmanian wolf (a carnivore).

Most of the mammals belong to subclass Theria, which contains the placental animals (those born in a relatively advanced state of development).

Subclass Theria. Therians are first known from rocks of Jurassic age, and constitute the largest group of mammals living today. They undergo considerable development before they are born, and at birth usually resemble their parents. This subclass has been divided into several orders, but only the more important ones will be discussed here.

Order Edentata. A rather primitive group of mammals, the edentates are represented by such living forms as the armadillos, tree sloths, and ant-eaters. Members of this order were common in North and South America in Pleistocene and Pliocene times, and fossil edentates are fairly common in certain American Cainozoic rocks. One such form was *Mylodon* (Fig. 145), one of the extinct giant ground sloths. These huge sloths were quite heavy, and some

Fig. 145. *Mylodon*, a giant ground sloth

of them stood as much as 15 feet tall. They were the forerunners of the modern tree sloths of South America.

The glyptodonts, another interesting group of curious edentates, were the ancestors of the present-day armadillos, and lived at about the same time as the ground sloths. *Glyptodon*, a typical Pleistocene glyptodont, was an armadillo-like beast with a solid turtle-like shell. In some forms the shell was four feet high. From the front of the bone-capped head to the tip of its tail a large individual might measure 15 feet long. There was a series of bony rings on the thick heavy tail, and in some species the end of the tail was modified into a bony, heavily spiked club.

Order Carnivora. The carnivores, or meat-eating mammals, are characterized by clawed feet and by teeth which are adapted for tearing and cutting flesh. The carnivores were first represented by an

Fig. 146. *Smilodon*, a sabre-toothed tiger

ancient group of animals called **creodonts,** a short-lived group that first appeared in the Palaeocene Epoch and was extinct by the end of the Eocene. The creodonts ranged from the size of a weasel to that of a large bear, and their claws were sharp and well developed.

These early meat-eaters were followed by more specialized carnivores which developed throughout Cainozoic time. Some examples are the sabre-toothed tiger *Smilodon* (Fig. 146), and the dire wolf *Canis dirus.*

Order Pantodonta. The pantodonts, known also as amblypods, were primitive, hoofed, plant-eating animals. They were distinguished by a heavy skeleton, short stout limbs and blunt spreading feet. The pantodonts appeared first during the Palaeocene and had become extinct by the end of Oligocene time.

Order Dinocerata. The members of this order are an extinct group of gigantic mammals commonly called uintatheres. *Uintatherium,* which is typical of the group, had three pairs of blunt horns, and the males had dagger-like upper tusks. Some of the uintatheres were as large as a small elephant and stood as much as seven feet tall at the shoulders. The size of the brain in relation to the size of the body suggests that these animals were not as intelligent as most mammals. Uintatheres are known from rocks ranging from Palaeocene to Eocene in age.

Order Proboscidea. The earliest proboscideans, the elephants and their relatives, were collected from Upper Eocene rocks of Africa. These early forms were about the size of a small modern elephant, but had larger heads and shorter trunks. Proboscidean development is marked by an increase in size, change in skull and tooth structure, and elongation of the trunk.

Two of the better known proboscideans are the mammoth and the mastodon, both of which inhabited Europe (including Britain), North America, Siberia, and regions as far south as Africa in Pleistocene times. The mastodons resembled elephants, but the

Fig. 147. Woolly mammoth, a Pleistocene mammal

structure of their teeth was quite different. In addition, the mastodon skull was lower than that of the elephant and the tusks were exceptionally large—some reaching a length of nine feet.

There were several types of mammoths, and the woolly mammoth is probably the best known of these. This animal lived until the end of the Pleistocene Epoch and, like the woolly rhinoceros discussed below, is known from ancient cave paintings and frozen remains. Information gathered from these sources indicates that this great beast had a long coat of black hair with a woolly undercoat (Fig. 147). Only two proboscideans exist today: the African and Indian elephants.

Order Perissodactyla. The perissodactyls (odd-toed animals) are mammals in which the central toe on each limb is greatly enlarged. The horses, rhinoceroses, and tapirs are typical living members of this order. Extinct members of the Perissodactyla include the titanotheres, chalicotheres, and baluchitheres, all of which grew to tremendous size and took on many unusual body forms.

Horses. The ancestor of the horse was *Hyracotherium* (also called *Eohippus*), one of the first perissodactyls. This small animal was about one foot high and its teeth indicate a diet of soft food. Following the first horse, there is a long series of fossil horses which provide much valuable information on the history of this important group of animals.

Titanotheres. This group of odd-toed mammals appeared first in the Eocene, at which time they were about the size of a sheep. By Middle Oligocene time, they had increased to gigantic proportions, but still had a small and primitive brain. *Brontotherium* bore a slight resemblance to the rhinoceros but was about eight feet tall at the shoulders. A large bony growth protruded from the skull and was extended into a flattened horn divided at the top.

Although the titanotheres underwent rapid development during the Early Tertiary, they became extinct during the middle of the Oligocene Epoch.

Chalicotheres. The chalicotheres were in some ways like the titanotheres, but they also exhibited many peculiarities of their own. The head and neck of *Moropus*, a typical chalicothere, were much like that of a horse, but the front legs were longer than the hind legs; the feet resembled those of a rhinoceros except that they bore long claws instead of hoofs. The chalicotheres existed from Miocene until Pleistocene time, but were probably never very numerous.

Rhinoceroses. The rhinoceroses are also odd-toed animals, and there are many interesting and well-known fossils in this group. The woolly rhinoceros was a Pleistocene two-horned form that ranged from southern France to north-eastern Siberia. It is well known from complete carcasses recovered from the frozen tundra of Siberia and from remains that were found preserved in an oil seep in Poland. These specimens, plus cave paintings made by early man, have given us a complete and accurate record of this creature. Fossil rhinoceroses have been found in rocks ranging from Middle Oligocene to Late Pliocene in age.

Baluchitherium, the largest land mammal known to science, was a titanic hornless rhinoceros that lived in Late Oligocene and Early Miocene time. These immense creatures measured approximately 25 feet from head to tail, stood almost 18 feet high at the shoulder, and must have weighed many tons. They appear to have been restricted to Central Asia.

Order Artiodactyla. The artiodactyls, or even-toed hoofed mammals, include such familiar forms as the pigs, camels, deer, goats, sheep and hippopotamuses. This is a large and varied group of animals, but the basic anatomical structure of the limbs and teeth reveals the relationship between the different forms. Artiodactyls are abundant as fossils in rocks ranging from Eocene to Pleistocene in age.

Entelodonts. These giant piglike artiodactyls lived during Oligocene and Early Miocene times, and were characterized by a long heavy skull that held a relatively small brain. The face was marked by large knobs beneath the eyes and on the underside of the lower jaw. The knobs, though blunt, had the appearance of short horns. Some of these giant pigs attained a height of six feet at the shoulders and had skulls three feet long.

Camels. The earliest known fossils of camels were collected from rocks of Late Eocene age. These were small forms, and the camels underwent considerable specialization of teeth and limbs as they developed in size. Many of the camels that lived during the Middle Cainozoic had long legs well adapted to running, and long necks which must have helped them to browse on the leaves of tall trees.

We have not considered such mammals as the rodents, including rats and mice, which were widespread from early Tertiary days; or those like the whale and dolphin which returned to the sea. The reason is that they are not very important as fossils. Neither have we discussed the Primates, the order which includes the apes and man: this will be dealt with separately in Chapter Nineteen (The Geologic History of Man).

H

CHAPTER SEVENTEEN

EVOLUTION: CHANGING LIFE

The earth's inhabitants have undergone continual and gradual change throughout the geologic past. The record of this change is well documented and may be explained by the theory of organic evolution. This theory, which is of paramount importance to both biologists and geologists, has been defined as a process of cumulative change characterized by the progressive development of plants and animals from more primitive ancestors. Thus, a study of evolution indicates that our modern-day plants and animals have attained their present degree of development as a result of gradual and orderly changes which have taken place in the geologic past. Because of its importance to the historical geologist, a brief review of the concept of evolution is appropriate here. (For more information on organic evolution, see *Biology Made Simple*, by Ethel M. Hanauer.)

(1) Theories of Evolution

The idea of organic evolution is not new, for certain of the early Greek philosophers had a rather crude concept of evolution as early as 500 B.C. It was not until the nineteenth century, however, that the first scientific theory was presented.

Although many scientists have studied the evolutionary changes undergone by plants and animals, relatively few have attempted to explain how these changes have come about. Among those who have are Jean Lamarck, Charles Darwin and Hugo de Vries.

Theory of Inheritance of Acquired Characteristics. Proposed by Jean Baptiste Lamarck in 1809, this theory contends that an organ which is constantly used becomes highly developed, and one which is not used becomes weak and eventually disappears. Lamarck was also of the opinion that these **acquired characteristics** could be transmitted from parent to offspring. Thus, after several generations, new or changed species would be produced. Lamarck believed, for example, that a blacksmith's son would be born with larger biceps than would the son of a shopkeeper, the difference being attributed to the superior physical condition of the blacksmith. Known also as the **Theory of Use and Disuse,** this theory has little support from the scientists of today.

Theory of Natural Selection. This theory was proposed in 1859 by Charles Darwin, one of the most famous of all biologists. Darwin, in his attempt to explain the cause of evolution, developed a theory based on a great deal of scientific research and experimentation.

Presented in his classic book, *The Origin of the Species*, Darwin based his theory of natural selection on four factors:

(*a*) **The struggle for existence.** All organisms produce far more offspring than can be expected to reach maturity. This overproduction results in a continual competition for food, water, shelter, and other necessities. The organism must overcome these difficulties or perish.

(*b*) **Variation.** No two of these offspring are alike, for there is variation even within an immediate family.

(*c*) **Natural selection.** Only those individuals that are best adapted to survive and reproduce their kind reach maturity. This leads to the 'survival of the fittest'.

(*d*) **Sexual selection.** Some individuals possess certain characteristics which give them an advantage in securing a mate, and these favourable characteristics are passed along to their offspring. Individuals lacking such attractive characteristics are unable to secure mates and would not, therefore, produce offspring.

Although modern scientists accept much of Darwin's theory, several objections have been raised to some of his ideas. As a result of these objections, certain aspects of Darwinism (as his theory is called) have been modified in the light of more recent scientific knowledge.

The Mutation Theory. Hugo de Vries, a Dutch botanist, proposed this theory in 1901. De Vries' work was based on the pioneer genetic studies of Gregor Johann Mendel, an Augustinian monk, whose original work was published in 1866. Mendel laid the foundations for the modern science of genetics.

The mutation theory is based on the belief that new species appear in nature as a result of mutations (sudden variations in the germ plasm of organisms) which are favourable for the survival of the organism. These helpful mutations may be inherited by offspring, and passed along from one generation to the next until a new species has been developed.

This theory, which compliments, rather than contradicts, the work of Darwin, suggests that new species may appear relatively suddenly, rather than as a result of minor changes over long periods of time.

(2) Evidence of Evolution

Although much of the evidence supporting evolution is indirect, it is gathered from many sources and is largely indisputable. For this reason, biologists and geologists consider evolution to be more factual than theoretical. Some of the more important types of evidence are briefly reviewed below.

Evidence from Comparative Anatomy. Many plants and animals which do not resemble each other, and appear to be totally unrelated, possess comparable anatomical structures. For example, the arm of a man, the wing of a bat, the flipper of a whale, and the wing of a bird, all possess basic structural similarities. They are, moreover, all organs of locomotion. Such structures or organs, which show fundamental structural similarities, are called **homologous structures.** They strongly suggest that the different species possessing them have a common ancestry, from which they have gradually diverged.

Many animals also possess structures or organs which, although recognizable as having once been useful, have no apparent function today. These are called **vestigial structures** and provide further proof of organic evolution. The vermiform appendix, for example, is a useless, if not actually troublesome, organ in man. But in certain other animals (such as the rabbit and dog), this appendix is still a useful part of the digestive system.

Evidence from Embryology. A study of an organism from the time of fertilization until its birth offers considerable evidence of the close relationship between the simple and complex forms of life. It is known that the early embryos of certain animals possess structures which resemble those of adult forms of less highly developed animals. Such structures may disappear as the embryo grows older, or they may remain as vestigial structures. For example, nonfunctional gill slits are present in all embryonic reptiles, amphibians, birds and mammals. Although they disappear before the birth of the animal, these primitive gills are seen as a relic of the past, and indicate a common aquatic ancestor for all of these forms.

The results of such embryological studies have given rise to the **biogenetic law,** or the **law of recapitulation.** This law states that 'ontogeny recapitulates phylogeny'. Or, more simply, the development of the individual (ontogeny) resembles closely, or repeats, the development of the race (phylogeny) to which the individual belongs.

Evidence from Classification. Taxonomy is based upon kinship of organisms. The Linnean system of animal classification begins with the simplest forms of animal life (the Protozoa) and proceeds to the most complicated (the Chordata). This classification, which is based on structural relationships, indicates a line of common descent which is most easily explained by evolutionary processes.

Evidence from Genetics. Studies in genetics (the science of heredity) have done much to favour the acceptance of evolution. Man, by means of artificial selection or controlled breeding, has been able to produce numerous varieties of plants and animals. For example, each of the many varieties of horses can be traced back to a single species of wild horse. Geneticists, in carefully controlled laboratory experiments, have been able to produce variations in organisms such as the fruit fly. Controlled breeding actually puts evolution on a practical basis. (It is true that no new *species* has been created by such breeding, but it should be remembered that man has been active in this respect for only 4,000 years, whereas in the evolution of a new species we are dealing with millions or tens of millions of years.)

Evidence from Geographic Distribution. The distributional relationships of certain animals are thought to be related to evolutionary change. There is evidence in some cases to suggest that species originally developed in certain central areas, but changed as they later became isolated. For example, the camel of Asia and the llama of South America had a common ancestry before the continents were completely separated. After the continents separated, evolution proceeded along two different lines. It appears then that isolation can lead to the development of new species.

Evidence from Palaeontology. Evidence gathered from studies of fossils provides one of the strongest arguments supporting organic evolution. Fossils actually show the progression of evolution because the oldest rocks bear fossils representing the simplest forms of life, and the fossil remains become increasingly complex in the younger rocks. Thus, when arranged in chronological order, fossils commonly show a progression that is most logically explained as the result of organic evolution.

Also, evolutionary trends can be seen within each group. For example: the graptolites show half a dozen trends, all of which were for the benefit of the colony; the micraster echinoids show well documented changes as they become more suited to a buried habitat.

Lastly there are in the phylum chordata (in particular) link fossils. We have seen earlier that it has proved difficult to decide whether some forms are reptiles or amphibians, and *Archaeopteryx*, the first feathered bird, has many reptilian features. Unfortunately there are still a lot of 'missing links' which blur the picture of evolution of our own species.

CHAPTER EIGHTEEN

EARTH HISTORY

The earth has undergone multitudinous changes during its long history. These changes, both physical and biological, had a marked effect on the climate, geography, topography, and life forms of prehistoric times. In this chapter we shall look at some of these changes and their role in geologic history.

(1) The Pre-Cambrian Eras

The Archaeozoic and Proterozoic Eras are commonly grouped together and referred to as Pre-Cambrian. Pre-Cambrian rocks have been greatly contorted and metamorphosed, and the record of this portion of earth history is most difficult to interpret.

The Archaeozoic and Proterozoic comprise the whole of geologic time from the beginning of earth history to the deposition of the earliest fossiliferous Cambrian strata. If the earth is as old as we believe it to be, Pre-Cambrian time may well represent as much as 85 per cent of all earth history.

Archaeozoic Era. The Archaeozoic Era covers a vast interval of time during which the earth appears to have been virtually devoid of life. There is, nevertheless, some indirect fossil evidence in the form of carbon-bearing deposits which may be organic in origin. However, most Archaeozoic rocks consist primarily of highly metamorphosed volcanic and sedimentary rocks, intruded by granite. Most are so greatly altered that they provide little information about their original nature. This was a time of igneous activity and mountain-building, marked at the end by a period of massive erosion.

Proterozoic Era. The rocks of the Proterozoic Era were formed after the long period of erosion which marked the end of Archaeozoic time.

This era is believed to have started more than 2,000 million years ago and included periods of glaciation, volcanic activity, and marine sedimentation. In general, Proterozoic strata contain more sedimentary material and less igneous and metamorphic rock than do those of the Archaeozoic.

Proterozoic rocks contain the oldest known direct fossil evidence of life. It consists largely of worm burrows, sponge spicules, radiolarians, and calcareous algae. Some of the marine algae secreted large masses of organic limestone and are the most abundant fossils of this time. There was apparently no life on the land. So far as is determinable from the poor fossil record, the climate of this time probably ranged from warm and moist to dry and cold.

Rocks of Proterozoic age contain some of the largest deposits of metallic ores known to man, providing such valuable materials as silver, gold, nickel, iron, copper, and cobalt.

(2) The Pre-Cambrian of Britain

The terminology used above is a fairly recent development, and the student will find that most text-books describe the rocks of both eras under the one heading 'Pre-Cambrian'. Other names that have been used for this vast era are Azoic and Eozoic, which are derived from the Greek for 'without life' and 'dawn of life' and also Archaeozoic.

Lewisian. The Lewisian rocks are nearly 3,000 million years old, and are the only definitely Archaeozoic rocks in Britain. Their occurrence is limited to the north-west Highlands of Scotland and to the Hebrides, where they are found notably on the island of Lewis. The Lewisian group contains a great variety of rocks which can be summed up as gneisses and schists, all metamorphic rocks. Intruded into them are a series of dykes, and the whole complex is heavily folded due to an ancient orogeny (disturbance).

Torridonian. The age of the Torridonian rocks is not known, but they rest directly on the Lewisian and have Lower Cambrian rocks above. The Torridonian is confined to Scotland, where, in Skye, it is some 20,000 feet thick. The Torridonian is a group of sedimentary rocks characterized by a red arkosic sandstone. Since arkosic rocks are rich in feldspar, which is a mineral prone to weathering in damp conditions, we may assume that they were rapidly deposited.

Moinian. The dating of the Moine Series is difficult because of subsequent folding, but is estimated at about 1,000 million years. The Moine is composed of an uninteresting series of metamorphic rocks, generally found thrust by a gigantic fault on top of the Torridonian and Lewisian.

Dalradian. Probably contemporary with or slightly younger than the Moinian, the Dalradian forms the Grampians and Central Highlands of Scotland. It consists of a well defined series of rocks, including greywackes, algal limestone and volcanic rock. In the uppermost beds some of the early Cambrian fauna are found, including trilobites, and it is thought that the upper Dalradian rocks mark the close of the Pre-Cambrian and the beginning of a new era.

These four groups are confined to Scotland. The small isolated outcrops in other parts of England and Wales are too difficult to date.

About 700 million years ago (i.e. towards the end of the Pre-Cambrian), Britain was part of a huge granite-crust 'continent' called Pangaea, which included Europe, the Americas and Africa. At the same time Australia was probably split in two by a narrow shallow sea. The Pre-Cambrian rocks form a stable granitic shield in central and western Australia.

(3) The Palaeozoic Era

The beginning of Palaeozoic time marks the beginning of the first accurate records in geologic history. Palaeozoic rocks have not been subjected to such great physical change as have those of the Pre-Cambrian, and there is an abundance of sedimentary rocks, many of which are quite fossiliferous.

The Palaeozoic Era, which began more than 600 million years ago, has been divided into six (in America seven) periods of geologic time. These periods were of varying duration, ranging from 20 million years to 100 million. The periods are separated on the basis of relatively brief, naturally occurring periods of broad continental uplift. During such times, the seas were drained from the continents. These periods of uplift were followed by times of submergence, during which portions of the continents were covered by the seas and were receiving sediments.

Let us now briefly review the periods of the Palaeozoic Era and learn something of their physical history, climate, and life forms.

(4) The Cambrian Period

The Cambrian, the oldest period of the Palaeozoic Era, is the earliest period in geologic history in which we find an abundance of well-preserved fossils. The period derives its name from *Cambria*, the Latin name for Wales. It was in Wales that these rocks were first studied.

The Cambrian started about 600 million years ago when much of Britain was submerged under a shallow sea. Slowly this sea became deeper as the region subsided, forming a vast Palaeozoic geosyncline which stretched from Ireland through Wales and the Grampians and on into Scandinavia. The base of the Lower Cambrian is a conglomerate followed by a series of unfossiliferous shales and grits, succeeded by the Lingulella group containing a rich fauna of trilobites (*Paradoxides*, *Agnostus*) and brachiopods (*Lingulella*). The earliest dendroid graptolite (*Dictyonema*) appeared in the Upper Cambrian. This group includes part of the Durness Limestone, whose lower beds are Cambrian, which rests directly on the Torridonian in Scotland. Some of the Welsh shales have metamorphosed to slates which are used as roofing tiles.

Probably the land mass that became America was much closer to Europe at this time and provided much of the material for the deposition of the above-mentioned rocks, although almost one-third was similarly submerged. Meanwhile the Australian sea narrowed and remained shallow.

Cambrian life was dominated by trilobites and inarticulate brachiopods. Trilobites were especially numerous, forming as much as 60 per cent of the total fauna. Present also were such invertebrates as protozoans, sponges, snails, worms, cystoids and dendroid graptolites. There is no record of terrestrial or fresh-water life, nor is there any evidence of the remains of vertebrates.

We can only speculate on the climate of the Cambrian Period. It appears, however, that climatic zones were not as clearly defined as they are today, and the overall climate was probably mild and equable.

(5) The Ordovician Period

The Ordovician Period, which lasted approximately 75 million years (25 million shorter than the Cambrian) was named from the Ordovices, an ancient Celtic tribe that once inhabited Wales.

The Ordovician rests unconformably on the Cambrian (in places on Pre-Cambrian), following a period of uplift. A renewed marine transgression initiated the system, which is divided into five series. Starting from the oldest, these are the Arenig, Llanvirn, Llandeilo, Caradoc and Ashgill. The last two are sometimes combined and called the Bala Series.

Essentially the rocks of the Ordovician were black graptolitic shales interspersed with shallow marine deposits containing a rich trilobite and brachiopod fauna. Some of the shales have been altered into slates. A feature of the geosyncline, which was not apparent in Cambrian times, was the activity of large volcanoes. Tremendous amounts of ash and lava were deposited, particularly in Snowdonia. Much of the material deposited at this time continued to come from the north-west, and the Scottish Highlands featured as a land mass. Two other low-lying land masses existed for some time during this period: the first stretched across the Irish Sea, the Isle of Man and into the Lake District; the second extended from South Yorkshire southwards to Kent and across to the Severn Estuary. Much work was done on the rocks and fauna of the Ordovician by Sir Roderick Murchison and Adam Sedgwick, and hence the names given to some of the fossils of that time: *Didymograptus murchisoni*, the 'tuning-fork' graptolite, and *Angelina sedgwicki*, a trilobite. In parts of the region an ancient coast-line complete with cliffs, stacks and other shore-line features has been located.

It must be remembered that the five series of the Ordovician are not always found complete in any one region. In parts of Shropshire, for example, we have Caradoc beds (Upper Ordovician) resting directly on Cambrian and Pre-Cambrian rocks.

Ordovician climates were probably mild and equable over much of the world, and climatic zones, if present, are assumed to have been much less marked than those of today.

The warm, widespread Ordovician seas must have been conducive to the expansion of the marine life of the time. This is suggested by the palaeontological record, which indicates that these seas probably teemed with sea-weeds, protozoans, brachiopods, bryozoans, corals, gastropods, lamellibranchs, cephalopods, trilobites, cystoids, blastoids, crinoids and the first echinoids and graptolites which featured forms with eight, four or two branches. Especially noteworthy was the development of exceptionally large, straight, cone-shaped cephalopods, some of which attained a length of 15 feet.

Although not found in Britain, the vertebrates were beginning to appear. The first were small, primitive, armoured fishes (Fig. 148), whose remains consist of fragmental bony plates and scales. These primitive back-boned animals are called ostracoderms, and their remains have been found primarily in the Rocky Mountains of America.

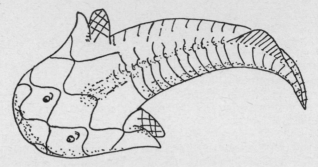

Fig. 148. Ostracoderm, a primitive jawless fish

(6) The Silurian Period

The Silurian Period also derived its name from an ancient Celtic tribe (the Silures) and was likewise first studied in Wales.

This period seems to have been of relatively short duration, about 20 million years (though some sources make it as much as 40 million). The division of the Silurian into series is still not settled, but most geologists recognize three series: the Llandovery, Wenlock and Ludlow. For a while the rocks deposited were similar to the Ordovician, but gradually the geosyncline silted up with masses of greywackes and graptolitic shales. On the margins of the trough we find limestones, with coral reefs, alternating with shales. The volcanic activity decreased, but other crustal disturbance was increasing as the Caledonian Orogeny came on. Except for the Highlands of Scotland, the seas gradually transgressed over the Ordovician lands, and two new 'fingers' of land appeared from the south-west towards the Severn Estuary and across southern Ireland (Fig. 149).

The uppermost bed of the Silurian, taken by some to be the base bed of the succeeding period, is remarkable. Known as the Ludlow Bone Bed and only inches in thickness, it contains a large number of fragments of the primitive fishes which up to then had been found only in America. There is evidence to suggest that they may have been fresh-water creatures and therefore may have occurred earlier without leaving a fossil record.

The **Caledonian Orogeny,** which was beginning during this period, reached its climax and turned the geosynclinal trough into a mountain region stretching across into Scandinavia. Associated with the tectonics was a burst of igneous activity, when granitic rocks were intruded into the older rocks. The intensely folded and faulted rocks that resulted have proved extremely difficult to unravel. Much work was done on them by Murchison and Sedgwick, but it was left to a school-teacher, Charles Lapworth, to sort them out. He used a technique based on that devised by William Smith, the 'Father of

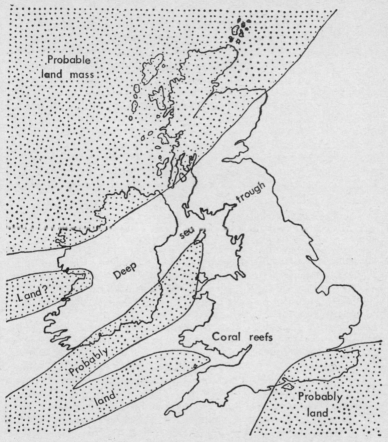

Fig. 149. The Silurian Period

British Geology', for correlating rocks according to their fossil content.

The end of the Silurian and the onset of the Caledonian Orogeny saw great changes in the fauna. The true graptolites became extinct and the trilobites declined; the vertebrates began to emerge; and new forms replaced older ones.

Silurian climates were probably warm and mild over wide areas. This assumption is based on the presence of large numbers of reef-building corals and the thick, widespread deposits of limestone and dolomite. The great salt and gypsum deposits of the Upper

Silurian suggest a period of pronounced aridity in those areas of
the world where such deposits are found.

The marine life of the Silurian is marked by an expansion of the
reef-building corals, articulate brachiopods, bryozoans, echinoderms, molluscs, graptolites (mainly *Monograptus* and complex multibranched and coiled forms). Trilobites reached their peak development before a rapid decline and near-extinction. Present also were
myriad invertebrates which had inhabited Cambrian and Ordovician
seas. Particulary distinctive were a group of scorpion-like arthropods
called eurypterids. These 'sea-scorpions' were the largest animals of
this time. The only vertebrates recorded from the Silurian are
primitive fishes similar to those of the Ordovician Period.

The earliest known land plants and animals appeared during the
Silurian. These early terrestrial animals were the scorpions and milli-
pedes which appeared in Late Silurian time. The fossilized fragments
of what appear to be land-dwelling plants have been found in Upper
Silurian rocks, principally in England and Australia. During
Ordovician and Silurian times the 'sea' in Australia narrowed and
started to migrate to the east.

(7) The Devonian Period

The Devonian is so named after Devonshire, where the marine
rocks were first studied. The period lasted for some 50 to 60 million
years. The Caledonian orogeny had created great mountain chains,
and most of Britain was part of a continent with a shore-line running across the southern counties. The period has therefore been
divided in accordance with the continental deposits: Old Red
Sandstone, and Marine Devonian.

Old Red Sandstone (often abbreviated to **O.R.S.**) outcrops in the
Midland Valley of Scotland, the Cheviots, and much of the English
Midlands and Wales. The beds contain few fossils and are typically
red, but include beds of various colours. The unfossiliferous limestones that occur are known as 'cornstones' and were probably
caused by evaporation, but typically the O.R.S. is composed of marls
and coarse sandstones. Volcanoes also exuded a great thickness of
ash and lava in the Midland Valley of Scotland. In southern England
a large delta and swampy region drained into the sea.

Marine Devonian. The marine beds which give rise to the red soils
of Devon are composed of a typical shelf-sea cycle of limestones,
shales and sandstones. During this period the sea transgressed and
retreated repeatedly as the region subsided. Nowadays a vertical
section passes through marine beds interspersed with terrestrial and
deltaic deposits.

Much of Devonian time appears to have enjoyed mild temperate
climates, and there is no evidence of marked climatic belts throughout the world. Devonian life was characterized by the spread of land
plants and the increase of the vertebrates. The plants included the
forerunners of the coal-swamp forest which were to dominate much
of the following period.

The invertebrates were represented by many species of reef-building corals, sponges, echinoderms (especially crinoids), lamellibranchs, and gastropods. Brachiopods were the dominant animals of this period; trilobites were present but still declining in numbers. A small cephalopod with a goniatitic suture pattern marked the first appearance of the ammonoids. Also appearing for the first time were the insects.

Vertebrates underwent unprecedented development, and both fresh-water and marine fishes were abundant. Among these were ostracoderms and the jawed, armoured placoderms. The true sharks appeared, as did a group of large sharklike forms known as arthrodires. Some of the arthrodires were heavily armoured and reached a length of 30 feet. However, the most important 'first' among the vertebrates was the appearance of the first tetrapod (four-footed vertebrate); this primitive amphibian appears to have evolved from the crossopterygians, or Devonian lungfishes.

(8) The Carboniferous Period

The Carboniferous Period lasted for over 65 million years and is subdivided into three distinct phases according to types of rock: Carboniferous Limestone; Millstone Grit; and Coal Measures. The Carboniferous Limestone is also known as the Lower Carboniferous Period and is roughly equivalent to what the Americans call the Mississippian Period. The Coal Measures make up the Upper Carboniferous Period, known in America as the Pennsylvanian Period.

Carboniferous Limestone. During this part of the period the sea steadily transgressed northwards over the remnant of the O.R.S. continent known as Saint George's Land, forming great thickness of limestones. These beds have been zoned principally on the coral forms included. As one travels northwards the base of the limestone is marked by successively younger zones (identified by the coral or brachiopod forms included in them). While this was happening a delta drained into central Scotland, which was involved in extensive volcanic activity. A sea also transgressed into Yorkshire across to the Pennines and produced, in the unstable conditions, a rhythmic cycle known as the Yoredale facies—basically a sequence of limestones, slates and sandstones, sometimes with coal bands.

Millstone Grit. The rocks which are known as the Millstone Grit once covered much of central England and were formed in deltaic conditions. The poorly sorted, false bedded and brecciated grits are interspersed with marine bands, a token of the unstable conditions that persisted. In the south a geosyncline extending through the English Channel was rapidly infilled with greywackes.

Coal Measures. Vast swamps, prolific growth, and continuing unstable conditions resulted in this unique phase of geologic history. As the sea-floor sank and the land rose, material was eroded and deposited and the balance upset, as a repeated rhythm of marine shales, lagoonal mudstones, deltaic grits, sandstones and mudstones and swampy fireclays and coals was set up. The distribution of coal-mining centres in Britain (see page 150) is an indication of the extent of the coal measures, which also continue out under the North Sea.

The Carboniferous Period was terminated by another great mountain-building phase known as the **Armorican (or Hercynian) Orogeny.** In the south the effects were most noticeable, as the Dartmoor granite was intruded with the accompanying mineralization and volcanic activity. Further north the effects were less pronounced, and the folds and faulting were gentler.

During the Carboniferous Period there was an abundance of life on land and in the water. The seas contained the usual hordes of invertebrates including foraminifers, bryozoans, brachiopods, cephalopods, blastoids, crinoids (which in places formed masses of crinoidal limestone) and corals. The brachiopods, which are used as zone fossils along with the corals, included the spiny *Productus*. The corals were used as zone fossils in the Carboniferous Limestone and are, from oldest to youngest: *Cleistopora* (K Zone), *Zaphrentis* (Z Zone), *Caninia* (C Zone). The next zone is named after the brachiopod *Seminula* (S Zone), which is often associated with the massive coral *Lithostrotion*. Last comes the *Dibunophyllum* (D Zone). These major zones have been subdivided still further. The lamellibranchs (pelecypods) advanced into fresh-water lakes. Insects made up an important part of the fauna, and included specimens with wing-spans nearly 30 inches across. Insects were necessary for the success-ful fertilization of some of the plants which were so prolific. The swamp forests were composed of large ferns and other plants; *Lepidodendron* and *Sigillaria* were prominent members of the flora.

Fishes continued to diversify, and the amphibians reached the peak of their development. The greatest step forward, as far as vertebrates were concerned, was the appearance of reptiles which had the ability to lay eggs with shells.

Meanwhile Australia had experienced a mountain-building epi-sode which virtually eliminated the 'sea' that had persisted since Cambrian days. Volcanic activity was a feature in the east, while, in the remnants of the sea, coral-rich limestones were deposited.

(9) The Permian Period

The Permian, named after the province of Perm in central Russia, is the last period of the Palaeozoic Era. It lasted about 50 million years. Permian climates, geography, faunas and floras were con-siderably different from those of preceding periods.

Following the Armorican orogeny Britain was largely an arid wasteland, and the Permian beds were poorly developed. The sand-stones that were deposited during this period and in the Triassic which followed it are often known as the New Red Sandstone. The most southerly occurrence of Permian rocks is in Torbay, Devon, where they are red sandstones and breccias. Further north, marls took over from the coal measures, and masses of sandstones were deposited in near-desert conditions with millet-seed sands and dune bedding.

An arm of the Zechstein Sea, a highly saline sea, extended into central and northern Britain where a unique deposit, the Magnesian Limestone, was formed. This dolomitic limestone included oolitic forms and concretionary 'cannon-ball' limestone in the form of spheres up to four inches in diameter. As the sea evaporated the economically important salt beds were formed.

The fossil evidence is scanty, but what little there is points to great changes. Trilobites which had already disappeared in Britain finally became extinct, the rugose corals disappeared, and the goniatitic ammonoids, the Palaeozoic crinoids and echinoids (except for the centrechinoids), the blastoids and the cystoids vanished for good. The brachiopods decreased in numbers and became small and stunted, but they survived. In the Tethys Sea, which lay roughly where the Mediterranean does today, new forms of life were evolving.

The picture was much the same in America, but not everywhere else. For the first time pronounced climatic changes occurred around the world. The Australian continent took on its present shape and dimensions, and, along with South Africa and South America, was heavily glaciated at the beginning of the period. Milder conditions gradually prevailed, and swamp lands comparable to those of the Carboniferous developed in much of Australia and Asia. Permian coals have been mined in Europe, India, China and Australia, and a few seams are found in America. The plants were much the same as those of the previous period. The amphibians and reptiles continued to dominate the land. Some reptiles were already showing distinct mammalian features.

(10) The Mesozoic Era

The Mesozoic, as its name implies, has been called the time of 'middle-life', because it represents the transition period from the relatively primitive plants and animals of the Palaeozoic to the more modern Cainozoic forms. During the 167 million years' duration of the Mesozoic Era there was an unprecedented expansion of land animals (especially reptiles). Also noteworthy was the appearance of the first mammals, flowering plants and birds.

(11) The Triassic Period

The Triassic, the earliest period of the Mesozoic Era, derives its name from the Greek word *trias,* meaning 'three'. This refers to the distinct threefold division displayed by Triassic rocks in central Germany where the system was first described.

Only the oldest and youngest series, the **Bunter** and **Keuper,** are present in Britain; the **Muschelkalk** does not reach our shores.

The **Bunter** includes mottled sandstone and pebble beds. In the north the Zechstein Sea was still evaporating, and the occurrence of pebbles called *Dreikanters* (see page 225) indicates the aeolian conditions. The **Keuper** is composed of sandstones, a marl (which is barely calcareous), and more salt deposits. Gradually the sea was approaching Britain's southern shores, bringing with it moister conditions, as indicated by the Tea Green Marls exposed in the Severn Estuary.

The Triassic Period ended with the formation of a group of passage beds known as the **Rhaetic Series.** For a long time they were considered part of the Jurassic system, but their fossil content is typically Triassic. Formed in lagoons cut off from the advancing sea by low barriers they are mainly clayey beds. The famous Rhaetic Bone Bed, though only about two inches thick, contains remnants of fishes, amphibians and reptiles. Limestones occur at the top of the series including the misnamed Cotham Marble, an ornamental limestone.

The presence of reptile and amphibian remains suggests mild climates for much of Triassic time. In some areas, the occurrence of fossil plants indicates warm humid climates. In others, thick deposits of salt and gypsum point to arid conditions.

Life in the Triassic was quite different from that of any of the preceding Palaeozoic periods. Many new groups appeared, both

Fig. 150. Belemnite

aquatic and terrestrial, and there were marked changes among the invertebrates, vertebrates and plants.

Triassic land plants were dominated by conifers, cycads and ferns. A few species of the typical coal plants were still living. The trunks of the fossilized trees found in the Petrified Forest in Arizona are the remains of some of the huge coniferous trees of the Triassic.

Marine invertebrates included large numbers of cephalopods, lamellibranchs, gastropods, echinoids, corals and representatives of most of the other invertebrate phyla. Brachiopods, though present, were much diminished in numbers. The ammonites (shelled cephalopods with frilled septa) were probably the most abundant and distinctive animals of this time. The belemnites (Fig. 150), relatives of the modern squid, were also abundant. Reef-building corals, similar to modern species, formed coral reefs in many parts of the world.

The vertebrates continued their rather rapid development and were of many different types. Sharks were still common in the seas, and the true bony fishes were increasing in number and variety. Amphibians, though now overshadowed by the rapidly evolving reptiles, continued to thrive.

The reptiles were the dominant vertebrates of Triassic time and were represented by the turtles, phytosaurs, marine reptiles and dinosaurs. Some of these species supply instances of a tendency for new forms to revert back to a former habitat. However, there is no reversal of evolution. The marine reptiles did not become fish again, but became adapted to marine life. Sharks (fish), ichthyosaurs (reptiles) and dolphins (mammals), all exhibit similar features which make them suited to their environment; such forms are said to exhibit **convergent evolution.**

The early dinosaurs were small when compared with the giant reptiles of the Jurassic and Cretaceous, but some of the marine reptiles attained great size. Among the latter were the streamlined, swordfish-like ichthyosaur and the clumsy plesiosaurs. Some of these reptiles were up to 40 feet in length, although the average was much

Fig. 151. Phytosaur

less. The semi-aquatic crocodile-like phytosaurs (Fig. 151) were another distinctive reptilian group of this period.

(12) The Jurassic Period

This period derives its name from the Jura Mountains between France and Switzerland. The Jurassic lasted for about 45 million years and is best known for the many reptiles which lived during this period. It is divided into Lower, Middle and Upper Jurassic.

The **Lower Jurassic** (also known as **Lias**) saw the sea, heralded by events in late Triassic time, transgressing over much of the country leaving offshore islands. Most of the rocks of this age are clays and shales, but limestones, including an iron-rich oolitic limestone, occur in the south of England. The sea-floor was unstable, and much work has been done on the series to plot the steady movement northwards of this warm muddy sea by studying ammonite zones. Two shales of the Upper Lias, the Alum and Jet shales, are worthy of mention. Alum was extracted commercially in the Whitby area, and jet is used in a small jewellery and ornament industry.

The **Middle Jurassic** includes the famous Great Oolite and Inferior Oolite: limestone deposited in shallow, clear, warm seas (as explained in Chapter Four). The warping of the sea-floor produced a geosyncline in miniature. Limestones also occur in other forms: gritty limestones known by their fossil content—for example. Trigonia

Grits (from a lamellibranch), and Clypeas Grit (from an echinoid); and the **Pea** Grit which contains larger ovoid 'ooliths'. Deltaic deposits are found around the margin of the land.

The **Upper Jurassic** commenced with a limestone and a coarse sandstone, the Cornbrash. Soon the seas became muddy and great thicknesses of clay, the Oxford and Kimeridge clays, were deposited. Between the Oxford clay (used extensively in brick manufacture) and the Kimeridge, clearer marine conditions were temporarily established, and more oolitic limestone and coral limestones were deposited. These beds are known as the Corallian. As the sea retreated southwards, southern England became a swampy forest region, with trees and plants different from those of the Carboniferous. Large reptiles which populated the swamps have left their footmarks in the mud. The last marine activity is to be seen in the Portland Beds, which include the last oolitic limestone and sands. The final beds, the Purbeck, are limestones which were deposited in fresh-water lagoons. It is here that we find a fossil forest, near Lulworth, and fossilized soils.

The Jurassic flora and fauna were abundant. The mild equable climates and epicontinental (shallow shelf) seas were conducive to flourishing life. Cycads were so numerous that the Jurassic has been called the 'Age of the cycads'. The Jurassic forests contained ferns, cycads, ginkgos, rushes and conifers.

Invertebrates were remarkably abundant. Reef-building corals, scleractina, and tabulate corals made their last reefs in the warm seas and around the lagoons. The lamellibranchs (pelecypods) flourished in marine and fresh-water forms. Echinoids diversified, and with them gastropods, forams, and bryozoans populated the seas. But by far the most important were the cephalopods, the ammonites and belemnites. The ammonites, with their rapid changes and complex suture patterns, zone much of the Jurassic rocks. (When the tide goes out they can be seen shining in the sun on the shale beach at Whitby, Yorkshire, where they are replaced by the golden iron pyrites. With them can be seen the pencil-shaped belemnites.) The arthropods, formerly represented in abundance by the trilobites, now included shrimps and crabs. The land, too, was invaded by snails and vast numbers of insects.

It was the reptiles—more specifically, the dinosaurs—that dominated the backboned animals. Fishes, largely primitive bony fishes and sharks, were numerous. Amphibians were not as abundant as in the Permian and Triassic, but their absence was compensated for by the great expansion of the reptiles.

On the land such quadrupedal plant-eating dinosaurs as *Brontosaurus* and *Diplodocus* attained lengths approaching 90 feet and weighed tens of tons. Present also were carnivores such as *Allosaurus*, a bipedal form about 35 feet long. *Stegosaurus*, an armoured dinosaur, was another distinctive Jurassic reptile.

This period also produced the first flying reptiles. Known as pterosaurs, these remarkable creatures had hollow bones and small light bodies. *Rhamphorhyncus*, a typical Jurassic form, had a long tail, sharp teeth, and a maximum wingspan of about two feet. Marine reptiles, such as ichthyosaurs and plesiosaurs, were still common in Jurassic seas.

An especially important event of this period was the appearance of the first primitive bird. Called *Archaeopteryx*, it is known from a feather, two skeletons, and the fragments of a third, found in the Solnhofen limestone quarries of Bavaria. *Archaeopteryx* still retained such reptilian characteristics as teeth, and had claws on its

Fig. 152. *Archaeopteryx*, a Jurassic bird

wings (Fig. 152). However, the feathers definitely establish it as a bird.

Undoubted mammal fossils are found in Jurassic rocks. Fragmental remains prove the existence of animals about the size of a large rat. The structure of their teeth suggests that some were herbivorous, while others were apparently meat-eaters.

(13) The Cretaceous Period

The Cretaceous Period, which lasted for approximately 70 million years, is characterized by thick deposits of white chalky limestone. In fact, the name Cretaceous is derived from the Latin word *creta*, meaning 'chalk'. The rocks of the Cretaceous were first studied in the white cliffs of Dover.

The eroded and peneplained Wealden anticlinorium has revealed a succession of clays and ferruginous (iron-containing) sandstones below the chalk. The lowest beds, the Hastings Beds and Weald

Clays, were laid down in similar conditions to those that prevailed at the end of the Jurassic. There is a slight unconformity between the Jurassic and Cretaceous due to tilting and a further retreat of the sea.

The deltaic swampland encouraged plant life and the continuation of the reptiles. Towards the end of the Lower Cretaceous the sea again began to make its presence felt, and a series of clays and the Lower Greensand were deposited. The Lower Greensand is today exposed in sandpits throughout the Weald, where it is seen as a browny-orange sand. This colour is due to the iron oxide limonite, which is the weathered product of glauconite. The presence of glauconite in the sandstone indicates that the deposits were marine.

Meanwhile, in Yorkshire, an arm of the continental sea was invading the land, bringing with it a completely different fauna from that which is found in the Weald beds of the same age. Here the lowest beds are the blue Speaton Clays which rest on the Kimeridge Clay (the Portland and Purbeckian of the Jurassic are absent). Gradually the sea extended southwards and the red chalk was deposited.

Most geologists nowadays regard the Upper Cretaceous as starting with the sticky blue Gault Clay which has a rich ammonite fauna. More Greensands and marls were deposited as the sea invaded southern England. Then the unique chalk was deposited and the two areas linked together. The conditions of the deposition of the chalk continue to give rise to discussion and argument. The estimated rate of deposition of these thick beds was about one foot per 30,000 years. This rate is associated with the formation of oozes in deep seas, but the chalk was probably formed in shallow seas. (The echinoids, ammonites, crinoids and other fauna are all shallow-marine creatures.) It is probable, too, that conditions were arid and rivers were very few. Therefore water-borne material did not pollute the seas, and the limestone chalk was able to build up. The Upper Chalk contains many bands of flint, a siliceous material.

The Cretaceous was marked by mild temperate climates, though they may have been somewhat cooler than those of the Jurassic. Evidence provided both by fossils (corals became rarer in Britain) and by sediments supports this climatic interpretation. There is, moreover, evidence of some early Cretaceous glaciation in Australia.

Plant life of the early Cretaceous consisted of cycads, conifers and ferns, and closely resembled Jurassic floras. The angiosperms, or flowering plants, appeared in the middle of the period, and by Late Cretaceous time the vegetation was quite similar to that of today.

The warm shallow Cretaceous seas teemed with a multitude of invertebrates and, as in the preceding periods of the Mesozoic, the molluscs were the dominant forms. Foraminifers with calcareous shells contributed greatly to the building of the chalk and are useful guide fossils for Cretaceous rocks. The *Micraster* echinoids, belemnites and crinoids are used to zone the chalk. Large numbers of lamellibranchs, gastropods, and cephalopods (particularly ammonites in great variety and abundance) were also present. The echinoderms were well represented by sea-urchins and heart-urchins and, to a lesser degree, by starfishes and 'brittle stars'.

The dinosaurs were the dominant vertebrates, but other reptiles, fishes, birds, and primitive mammals were quite abundant. The fishes were similar to our present-day forms and are well known from their fossils. Birds were more specialized and more abundant than they had been in the past, and they have left an interesting fossil record. *Hesperornis*, a large flightless bird, is typical. Cretaceous mammals were small and possibly resembled the modern shrews or hedgehogs.

The reptiles continued to rule land, sea and air, and developed more striking forms than ever before. The ornithopods were well represented by such genera as *Trachodon* (known also as *Anatosaurus*) and similar duck-billed dinosaurs. The armoured ankylosaurs were quadrupedal herbivorous forms that lived only during Cretaceous time. *Ankylosaurus* is typical of this group. Even more distinctive were the ceratopsians, a group of horned dinosaurs. *Triceratops*, the largest of this group, was a heavy-set quadrupedal plant-eater, as much as 30 feet long. Its eight-foot skull was characterized by a parrot-like bill, and a heavy, frilled, bony shield which protected the back of the neck. *Tyrannosaurus*, the largest of the flesh-eating dinosaurs, stood 20 feet tall, was 40–50 feet long, and weighed many tons. Its forelimbs, greatly reduced in size, were armed with sharp claws suitable for grasping and tearing prey. Its giant skull was armed with powerful jaws and dagger-like teeth—some as much as six inches long.

The flying reptiles had continued to evolve, and by Cretaceous time some new and unusual forms were present. One of the most interesting pterosaurs was *Pteranodon*. This lightweight, hollow-boned animal had a body little more than two feet long, but a wing-span of as much as 25 feet. Its tail was short but the back of the head was marked by a long triangular extension of the skull.

Cretaceous seas supported a rare assortment of marine reptiles. Ichthyosaurs and plesiosaurs were still present, and they had been joined by the mosasaurs—giant lizard-like swimming reptiles (Fig. 153). These huge creatures, some as much as 50 feet long, had a typical lizard-like body, a flattened tail, and large, sharp, recurved teeth, and all four limbs were modified into flippers. The giant sea-turtles were another interesting group. Some, such as *Archelon*, were as much as 11 feet long and 12 feet across the flippers.

What happened to the dinosaurs? Why did they suddenly (geologically speaking) become extinct? This is a question that has long perplexed both biologists and palaeontologists. They did not become extinct because they were unsuccessful, for they lived for over 100 million years, attained great variety in size and shape, and successfully invaded any number of different environments. As to why, we can only speculate. Some ideas that have been advanced are: (*a*) the climatic and geographic changes of the Cretaceous were too drastic and the dinosaurs could not adapt to them; (*b*) the group was in 'racial old age' and their 'time was up'; (*c*) a widespread

Fig. 153. Swimming Mesozoic reptiles: (*a*) *Ichthyosaur*, (*b*) *Mososaur*, (*c*) *Plesiosaur*

epidemic or plague wiped them out (there is little, if any, scientific support for this); (*d*) the mammals, which were rapidly increasing in numbers, may have been eating the dinosaur eggs; and (*e*) changes in the Cretaceous flora were unacceptable to herbivorous forms for they could not eat them. As the plant-eaters disappeared, the carnivores which preyed on them died out also.

Most scientists believe that dinosaur extinction was a result of not one, but several of the above reasons. We shall, in all probability, never really know the answer to this baffling question.

It was probably at the end of the Cretaceous that much of the Continental drift took place. North America separated from Europe, South America from Africa; Antarctica, Australia and Asia became separate units. The enormous lakes and coal-forming swamps of the Jurassic were flooded by a sea which split Australia in two. Climatic conditions deteriorated and the Ice Age started to have its effects. The separations of land from land were to have remarkable effects on the fauna. The monkeys that evolved took different lines in the Old and New World. A few remnants of the dinosaur reptiles have hung on to the present day in the islands. Monotremes and marsupials have persisted locally, and so have flightless birds such as the emu. The amphibious frogs of Australia and South America are unique. The Australian snakes are different from those of the rest of the world, and are not found in New Zealand at all. But these are all developments which occurred in the Cainozoic, and it is to this era that we must now turn.

(14) The Cainozoic Era

The Cainozoic (often spelt Cenozoic) Era lasted for about 63 million years—a relatively short time when compared with the preceding geologic eras. The word Cainozoic literally means 'recent life' and refers to the large numbers of recent plant and animal species which developed during this time.

(15) The Tertiary Period

The Tertiary Period, which lasted for about 62 million years, derived its name from an early and now unused classification of rocks. The period has five well-defined epochs. These epochs, with the oldest listed last, and the literal translation of their names, are:

Pliocene—more recent
Miocene—less recent
Oligocene—little recent
Eocene—dawn recent
Palaeocene—ancient recent

The **Palaeocene** is not evident in Britain, and the oldest Tertiary beds are those of **Eocene** age which outcrop in two large basins in the south of England: the London Basin and the Hampshire Basin. The two basins are fairly similar, but there are differences between beds of the same age. Clays and sandstones are predominant and include lignites. The Eocene beds are typically revealed in the cliffs of White-cliff Bay in the Isle of Wight, and it is clear that a cycle of alternate marine planes and continental conditions prevailed. Along with the extinction of the great reptiles, the ammonites and belemnites disappeared and other great changes occurred; the development of grasses encouraged that of hoofed mammals. However, because of the marked similarity between Tertiary and Quaternary organisms, the life forms of these two periods are discussed together at the end of the chapter.

Meanwhile Scotland was subject to vast volcanic activity. A belt of volcanoes, extending into Ireland and across to Iceland, produced the great lava flows seen today in the Giant's Causeway and Fingal's Cave. Dyke swarms and massive parallel east–west dykes are further evidence of igneous activity.

The **Oligocene** in Britain is only found in the Hampshire Basin with clays and limestones,which contain corals, including **Madrepora,** a Hexacoral (Scleractina). Many of the beds were not marine, being deposited in brackish and fresh-water conditions. Fresh-water mussels are common, and remains of birds, mammals and reptiles have been found.

Rocks of **Miocene** age do not occur in Britain. This episode corresponds to the **Alpine Orogeny** which formed the Alps, the Himalayas, the Pyrenees, and some mountains in America. Britain was on the fringe of the region affected, the results being gentle east–west folds (mere ripples), such as the Weald Dome, and associated faults.

The **Pliocene,** although forming crags in East Anglia, was more important as a period of erosion. Peneplaination was the order of the day, a cycle of levelling off by erosion, interrupted by rivers cutting valleys and producing a hilly landscape. Subsequent further erosion is seen at various levels.

(16) The Quaternary Period

The Quaternary, like the Tertiary, derives its name from an early, outdated rock classification. A relatively short geologic period (about one million years), the Quaternary is divided into the Pleistocene Epoch and the Recent Epoch.

The **Pleistocene,** known also as the **Great Ice Age,** is characterized by four major glacial periods and three intervening warmer periods during which the ice melted. Great sheets of ice in northern Europe, Siberia and North America brought about colder climates which hastened the extinction of many Tertiary plants and animals. These great glaciers spread as far south as a line joining the Thames and Severn estuaries.

Such great masses of ice had a marked effect on the earth and were responsible for fluctuations of the sea level, depression of the land, and considerable change in the drainage pattern of rivers. The glaciers removed countless tons of soil and scoured the surface of the bedrock. In addition, Pleistocene glaciation is largely responsible for the formation of innumerable lakes and tarns. The boundary between the Pliocene and Pleistocene is indefinite, but the fauna of the fresh-water crags that were still being deposited changed from warm-sea types to those associated with an Arctic climate as the colder conditions prevailed. The evidence of the action of the glaciers was discussed in Chapter Eight. Much of the landscape of northern England, Wales and Scotland was determined during this time.

Though not very definite, the boundary between the Pleistocene and **Recent** is usually considered as the time when the last ice sheet retreated from Europe and North America—some time between 12,000 and 15,000 years ago.

Cainozoic life was characterized from an early stage by faunas and floras which in many respects were quite similar to those of today. The marine life closely resembled that of today, and the mammals were rising so greatly in importance that they were on the way to ruling the land. Plants were essentially modern in appearance. Forests of hardwoods and grassy plains provided suitable environments for the mammalian expansion that was taking place.

Invertebrate faunas, though resembling their Cretaceous forebears, were decidedly modern in aspect. Foraminifera, which were present in great numbers, are valuable guide fossils of the Tertiary—especially for the petroleum geologist. Corals, bryozoans, echinoderms (particularly echinoids) and arthropods were also abundant (corals became rarer, but are still found in British waters today). Brachiopods, which had so thoroughly dominated early Palaeozoic seas, were now greatly diminished in numbers and variety. The molluscs were still the dominant marine invertebrates. The ammonoids, so common throughout Mesozoic time, were replaced by an unprecedented number of highly diversified lamellibranchs and gastropods. Many of these species were quite like the clams, oysters, and snails of today.

Tertiary vertebrates are also well known, for there are fossilized remains of fishes, amphibians, reptiles, birds, and, to a greater extent, mammals. Fishes of the Tertiary were plentiful and included many bony fishes as well as large numbers of sharks. Some of the latter were 60–80 feet long and had six-inch teeth. The amphibians were represented by salamanders, toads and frogs. The reptilian hordes of the Mesozoic had dwindled to snakes, lizards, crocodiles, and turtles, which were present in about the same numbers as their modern descendants.

The majority of Tertiary birds were also much like those of today. Unfortunately, however, because of the fragility of their bodies, they are not often found as fossils. Of particular interest are the so-called 'giant' birds of the Tertiary. Some of these great ostrich-like flightless creatures were ten feet tall and laid eggs over a foot long. *Dinornis* and *Diatryma* (Fig. 154) are typical.

We have already mentioned the isolation of various parts of the world, including Australia, and the distribution of reptiles, monotremes, marsupials and flightless birds that resulted. Australasian climatic conditions were vastly different from those of the northern hemisphere. The Ice Age was coming on in the Cretaceous, and the last of the waves of glaciation occurred at the beginning of the Pleistocene when Europe was only beginning to suffer from the invading ice. Increasingly the scene became one of aridity, with its accompanying wind erosion, dried-up salt lakes, and stark barrenness. (The snowy peaks of New Zealand and cloud forests of New

Fig. 154. *Diatryma*, flightless Eocene bird, about seven feet tall

Guinea were exceptions to this trend.) Sand dunes, coarse grass and scrub dominated the land, apart from the woodlands around Perth with their famous Eucalyptus trees.

The greatest Tertiary development was, of course, among the mammals. Those of the Palaeocene Epoch were primitive and small, and not yet very similar to those of today. Eocene forms were somewhat larger and included the earliest rodents, camels and rhinoceroses. *Hyracotherium* (also called *Eohippus* the 'dawn horse', see Fig. 155), the earliest known horse, also made its appearance during the Eocene. Another group originating during this epoch was the creodonts, forerunners of the carnivores.

During Oligocene time, the mammals took on a more modern appearance. These more advanced types included dogs, cats, camels, horses, rhinoceroses, pigs, rabbits, squirrels and (in Africa) small elephants.

Also living during the Eocene and Oligocene were certain strange mammals quite unlike any that are living today. These included the *Dinocerata* or uintatheres, great rhinoceros-like beasts, some of which stood seven feet tall at the shoulder. The titanotheres, another group of gigantic early Cainozoic mammals, appeared first in the Eocene. Although originally about the size of sheep, they had increased to titanic proportions by Middle Oligocene time.

Mammals became even more varied and abundant in the Miocene—so much so that this epoch has been called the 'Golden age

Fig. 155. *Hyracotherium* (*Eohippus*), an Eocene horse about the size of a fox

of mammals'. This accelerated mammalian development was due in no small part to the spread of the grasses which blanketed Miocene plains and prairies. Horses, camels, deer, pigs, and rhinoceroses were amongst the familiar mammals that were present in various parts of the world. The Pleistocene also had its share of strange, now-extinct, mammals. Especially worthy of mention is the giant hornless rhinoceros called *Baluchitherium*. This tremendous creature, the largest land-dwelling mammal of all time, was 30 feet long and stood 18 feet high at the shoulders. The baluchitheres first appeared in the Oligocene and became extinct during Miocene time. These great beasts seem to have been restricted to Asia, for their remains have not been found elsewhere. Another interesting Oligocene–Miocene form was the entelodont, or giant pigs. Some of these creatures were six feet high at the shoulders.

Pliocene mammals were even more highly developed than those of the early Tertiary epochs. Giant ground sloths, such as *Mylodon*, were common in the southern United States during Pliocene and Pleistocene times. Some were 15 feet tall and weighed thousands of pounds. The glyptodonts, distant relatives of the armadillos, also evolved in the Americas.

During the Pleistocene the proboscideans (elephants and their kin) underwent considerable development. The mastodon and woolly mammoths were quite common in North America and Siberia, where they have been preserved in ice. They are also known to have lived in the British Isles. The woolly rhinoceros is known from remains in America and from the pitch deposits of Poland. Carnivores were well represented by *Smilodon*, the sabre-toothed tiger, and

Fig. 156. *Canis dirus*, the dire wolf

Canis dirus, the dire wolf (Fig. 156). *Smilodon*, which was about the size of a lion, had powerful jaws and highly developed, dagger-like upper canine teeth.

Probably the most noteworthy of all Pleistocene events was the appearance of the earliest known men. The evolution of the primates and the geologic history of man will be the topic of the next chapter.

CHAPTER NINETEEN

THE GEOLOGIC HISTORY OF MAN

Man is a relative newcomer on the geologic scene, having appeared some 600,000 years ago during the Pleistocene Epoch, or Great Ice Age.

The Primates, the mammalian order to which man belongs, have well-developed brains, elongated limbs, and cupped or flattened nails on their flexible fingers and toes. Other members of this highly developed order are the lemurs, tarsiers, monkeys, and apes. Man, along with the monkeys and apes, has been assigned to the suborder Anthropoidea. Members of this group are characterized by large brains and large eyes that face forward.

The fossil record of the Primates is not, unfortunately, as complete as the palaeontologist would like it to be. Nevertheless, with the aid of increased palaeontological and anthropological research, the geologic history of man is slowly being pieced together.

(1) The First Primates

The remains of what is reputedly the earliest known primate were found in the United States in rocks of Palaeocene age. These bones represent a small animal which apparently resembled the present-day lemur. Another primitive lemur, *Notharctus*, lived in North America during Tertiary time, for its remains have been found in certain Eocene formations of the western United States. *Notharctus* was a long-tailed, small-faced, tree-dwelling animal not unlike our modern lemurs. However, the present-day lemurs no longer inhabit North America, but are found largely in Madagascar. A few are found in Africa and Indonesia.

The tarsiers, rather delicate, wide-eyed creatures about the size of a rat, also appeared during the Eocene. These animals live in the jungles of Indonesia and the Philippines today, but their fossil record indicates that they inhabited North America and Europe during Eocene time.

The monkeys and apes made their initial appearance in early Oligocene time. Remains of the earliest monkey, *Parapithecus*, were collected in Egypt, as were the bones of *Propliopithecus*, the oldest known ape. Another important fossil ape was found in the Miocene of Africa. This ape, named *Proconsul*, has certain anatomical characteristics which suggest that it may have been ancestral to the chimpanzee, gorilla and man.

(2) The Manlike Apes

The remains of a group of manlike apes (known also as **australopithecines** or southern apes) have been found in Pleistocene cave deposits of South Africa. The first of these, *Australopithecus africanus*, was discovered in 1925 and consisted of the incomplete skull of a child about five years old. The character of these and other skeletal fragments discovered later suggests a close structural relationship with early man. However, scientists disagree on the exact evolutionary position of these primitive creatures. Some believe them to be the earliest known men, while others think they were a terminal group of manlike apes and are not, therefore, ancestral to man. Except for his relatively small brain capacity, *Australopithecus* was a fairly advanced creature. He had an almost upright posture, was bipedal, stood about five feet tall, and was characterized by an apelike head, powerful jaws, and teeth closely resembling our own.

(3) From Prehistoric to Modern Man

The earliest human (or near-human) remains have been found in rocks of Early Pleistocene age in Africa. A succession of Middle and Late Pleistocene humans follow this early man, and these primitive people are known from both skeletal remains and artefacts (implements or weapons of prehistoric age).

Most of these early fossil men have been given a name that refers to the geographic locality in which their remains were first discovered. Thus, Peking Man (*Sinanthropus pekinensis*, meaning 'Chinaman of Peking') refers to a prehistoric man who lived in the vicinity of Peking (now Peiping), China.

East Africa Man. The remains of what appears to be the oldest known human were found in rocks of early Pleistocene age in Oldoway (Olduvai) Gorge, Tanzania, East Africa. Consisting of a lower jaw, two skull bones, the foot bones, a collarbone, and some of the hand bones, these fossils are believed to represent the remains of an 11- or 12-year-old child. The bones of the child were found in association with the remains of another person thought to be an adult.

These bones, estimated to be more than 600,000 years old, are older than *Zinjanthropus boisei* (the so-called East Africa Man) discovered at Olduvai Gorge in 1959. The original discovery, a skull, is believed to have belonged to an 18-year-old boy. All of these remains were discovered by the field party of Dr. Louis S. B. Leakey, formerly Director of the Kenya National Museum, Nairobi. With these ancient skeletons, Dr. Leakey and his group found a variety of crude pebble tools and the remains of extinct Pleistocene animals.

Dr. Leakey's son, Richard, has switched attention from Olduvai to Lake Rudolf in the northern desert region of Kenya where he described the 1,000 square miles of sedimentary lake deposits as a natural museum of fossils. Fragments of some 100 individuals have been discovered including the controversial 'KNM-ER 1470' (Kenya National Museum—East Rudolf) skull found in July 1972. This, together with subsequent finds of a dozen or so *Homo* individuals, was found in the same beds as *Australopithecus*, suggesting that *Homo sapiens* is not, as his father believed, an off-shoot of the Australopithecine line. These discoveries have caused considerable rethinking and controversy. Further finds will perhaps settle this question.

Java Ape Man. The remains of this primitive manlike creature were first collected in 1891 near the village of Trinil in Java. Officially named *Pithecanthropus erectus*, the erect ape man, Java Man probably lived 400,000 to 500,000 years ago. His bones have been found with the remains of extinct Pleistocene elephants, rhinoceroses and tapirs.

A reconstruction of *Pithecanthropus* suggests an individual about 5½ feet tall, with a broad apelike skull marked by a sloping forehead, flat nose, and chinless jaw. The teeth, however, are near-human, and the brain capacity is larger than that of the average adult ape (Fig. 157*a*).

Peking Man. *Sinanthropus pekinensis* is known from the fossil remains of at least 40 individuals who once inhabited the area of

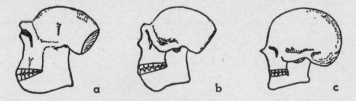

Fig. 157. Skulls of fossil men: (*a*) Java Ape Man, (*b*) Neanderthal Man, (*c*) Cro-Magnon Man

Peking in China. With the exception of a larger brain capacity, Peking Man is in many ways physically similar to Java Man. Some authorities have even suggested that Peking Man should be assigned to the same genus as Java Man—he would then be known as *Pithecanthropus pekinensis* rather than *Sinanthropus pekinensis*. However, other anthropologists and palaeontologists believe Peking Man to be considerably more advanced, as he apparently made crudely fashioned stone tools and knew how to use and control fire.

There is a current trend among certain zoologists to put all fossil men in the genus *Homo* (see below). Java Man would thus be known as *Homo erectus*, Peking Man as *Homo erectus pekinensis* (a subspecies of Java Man).

Heidelberg Man. The oldest European human fossil, *Homo heidelbergensis*, consists of a pair of lower jaws and 16 well-preserved teeth. This creature, called Heidelberg Man (the remains were found near Heidelberg in southern Germany), probably lived about 450,000 years ago and may have been intermediate between man and the apes. It has also been suggested that Heidelberg Man may have been a close ancestor of Neanderthal Man (see below).

Neanderthal Man. Probably the best known of all fossil men, *Homo neanderthalensis* was widespread in Europe and Asia during the Late Pleistocene. The first neanderthaloids were found in 1856 in the Neanderthal (Neandertal) Valley of north-western Germany, but were not recognized as a distinct human species until 1864.

The large number of neanderthaloid remains has provided the palaeontologist with a fairly clear picture of the physical characteristics of this early group. The typical individual was approximately five feet tall, with stooped shoulders, and the knees were slightly bent. This tended to give the body a slouched appearance. The head was large (the brain capacity approximates that of present-day man), and was characterized by a flat nose, receding chin and a low forehead with heavy brow ridges (Fig. 157*b*).

Neanderthal Man appears to have been a cave dweller (hence the name 'cave-man'); he made well-formed stone implements and knew how to kindle and use a fire. There is also evidence to indicate that he buried his dead.

Although **Piltdown Man** no longer has scientific status, it is only fitting that 'he' be included in our discussion of prehistoric men. This fossil 'man' was the centre of one of the greatest, and most successful, scientific hoaxes ever perpetrated. The remains, which were given the scientific name *Eoanthropus dawsoni*, were collected from Pleistocene gravel deposits near Piltdown in Sussex. The skull was definitely human, but the lower jaw and teeth were apelike in appearance. Many scientists were suspicious of the great difference between the jaw and the skull, and also of the conditions under which the fossil was found. Finally, after 40 years of intensive investigation, the Piltdown 'fossils' were proved to be the carefully stained and abraded modern skull fragments of a human, and the lower jaw of an orang-utan (an anthropoid ape).

Modern Man. The earliest known modern man or *Homo sapiens* (the species to which you and I belong) appeared about 35,000 years ago. These early men, called **Cro-Magnon** because their remains were first discovered in 1868 at the rock shelter of Cro-Magnon in the Dordogne Valley of France, are known from a large number of well-preserved skeletons. They were a well-built, rugged people (many of the skeletons exceed six feet in height), who walked erect and had an essentially modern skull with a well-developed chin, pointed nose and high forehead (Fig. 157*c*). Cro-Magnon Man's well-developed tools fashioned from stone, bone and horn, and his artistic talent, as seen in paintings found in many of his caves, give evidence of the superior mental development of the group.

So man evolved and *Homo Sapiens* arrived in Britain. The oldest remains were found at Swanscombe in Kent and date from the second interglacial period, but the real invasion came with the end of the Ice Age when men lived in caves. The rest of man's story belongs to the science of anthropology, to archaeology and later still to history.

CHAPTER TWENTY

GEOLOGICAL MAPS

The student of geography is used to maps and soon learns to read them and draw conclusions from them, even about parts of the world he has never seen. Similarly, the student of geology derives a wealth of information from his geological maps. Map reading is in fact an important part of geological studies at all levels. Therefore this last chapter is included as a summary of what has gone before, and as an introduction to the geological map.

The first geological maps were drawn by William Smith in 1815. By observing the rocks and correlating them by the use of fossils, he was able to produce a series of 15 maps (scale five miles to the inch) which covered England and Wales. Today the Institute of Geological Sciences publishes a series (one mile to the inch) which covers most of Britain.

The situation in the field is often complicated, and the student is advised to consider problem maps first. These maps are devised to show particular aspects of geology in an imaginary area. Let us look at one such map (Fig. 158) to see how it reflects the information given in the first part of this book.

First, let us see what kinds of rocks are present.

Metamorphic rocks: schist and ocal lhornfels around intrusion
Igneous rocks: dolerite dyke and basalt
Sedimentary rocks: sandstone, limestone, grit and arkose
Other features: raised beach, river terraces and glacial striae

Secondly, let us consider the structures in this area.

In the north we have the basalt, which is horizontal. (Notice how the base of a horizontal layer follows the contour lines.)

Farther south we have the grit and arkose dipping at 5°N. There is therefore an unconformity between these and the basalt. Notice how dipping rocks form Vs in valleys and on hills. As we continue southwards, we find sandstone and limestone with a fault running east–west and an anticline and syncline parallel to it. The northern-sloping limbs dip at 30° and the southern-sloping limb dips at 20°; the fold is therefore asymmetrical. There must also be an unconformity between these rocks and the grit and arkose. (Notice how the fault runs straight across country: this means that it is vertical, and that folds produce $\times$ and $\subset$ patterns as the two dipping limbs meet.)

In the east we have a dyke with a locally metamorphosed zone. As this runs straight, it too is vertical. Lastly, outcropping in several locations is a schist which is a regionally metamorphosed rock, so

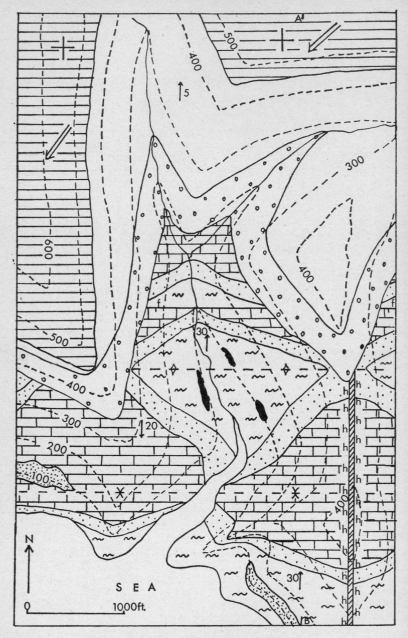

Fig. 158. Problem map referred to in text

KEY

 Basaltic lava,

 Grit

Arkose

Sandstone

Limestone(with corals and crinoids)

Schist

Dolerite & Hornfels

Raised beach

River terrace

Horizontal rock

Glacial striation

Rock boundary

34 → Direction & angle of Dip

—◇— Axis of anticline

—✗— Axis of syncline

f Fault (tick on down-thrown side)

--200-- Contours at one hundred foot intervals

Fig. 159. Key to Fig. 158

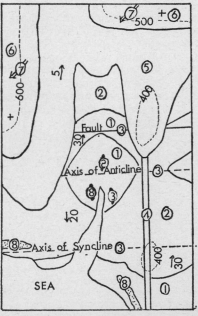

Fig. 160. Diagram of order of events in Fig. 158

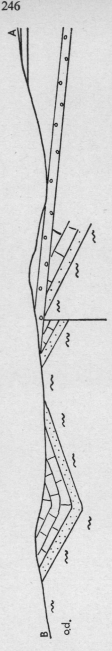

Fig. 161. Section
A-B

there is a third unconformity between this and the sandstone and limestone.

Thirdly, let us put the events in order (Fig. 160). The schist is the oldest rock, and on this rest the folded and faulted strata of sandstone and limestone. The dyke and hornfels must have come later, being intruded into the schist, sandstone and limestone, but they are overlain by the other rocks. The fourth group comprises the gently dipping grit and arkose sequence and then the horizontal basalt. Lastly we have the recent glacial striations, river terraces and raised beach.

Fig. 161 is a section taken along the line AB. This is an interpretation of the events outlined above to help us build up a three-dimensional view of what has happened.

Now let us reconstruct the geological history of the region, using the information that we have gleaned above, and our knowledge of geology from previous chapters of this book.

The first event recorded in this region is the alteration, on a regional scale, of some previous rock to form the schist. This probably occurred under intense pressure and heat. There then followed a period of erosion and marine transgression. This built up about 100 feet of sandstone and several hundred feet of limestone (thicknesses can be obtained from the section). The change from sandstone to limestone occurred either as the sea deepened or as the amount of material entering the sea decreased. Corals and crinoids have been found in the limestone, a fact which indicates that the seas were shallow, warm and clear. There may have been further beds in this sequence, but they have been lost in the uplift, folding, faulting and erosion that followed. The normal fault produced a downthrow of 300 feet to the south (this too can be obtained from the section). A period of igneous activity took place at this time, as evidenced by the dolerite dyke, which also altered the country rocks by heat of intrusion.

After the period of erosion, the grit and arkose were deposited. Since an arkose is a sandstone with a high feldspar content, these rocks must have been deposited in arid conditions similar to those prevailing in Britain during Permian and Triassic times.

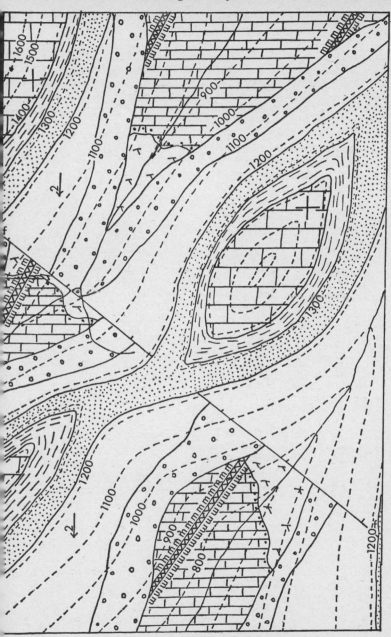

Fig. 162. A second problem map

A second spell of igneous activity then followed. We have two parts of a basalt lava flow outcropping on higher land in the area. (Notice that the resistant dyke also forms a ridge of low-lying hills). So, following the arid conditions, volcanoes extruded a thickness of at least 200 feet of lava, the underlying beds having been tilted.

The hot arid conditions vanished completely and the region was later covered by a glacier. It is possible to determine the direction of movement of this glacier from the marks it has left on the lava flows. In recent days a river has eroded itself a valley. River terraces and the raised beaches indicate a period of uplift when the river was rejuvenated, forming the present-day topography.

A second map (Fig. 162) is included for the reader to try and interpret for himself.

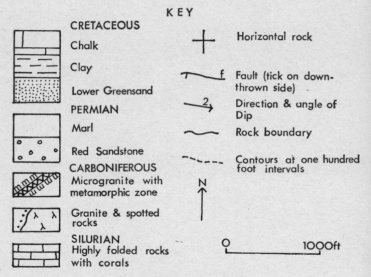

KEY

CRETACEOUS
Chalk
Clay
Lower Greensand
PERMIAN
Marl
Red Sandstone
CARBONIFEROUS
Microgranite with metamorphic zone
Granite & spotted rocks
SILURIAN
Highly folded rocks with corals

Horizontal rock

Fault (tick on down-thrown side)

Direction & angle of Dip

Rock boundary

Contours at one hundred foot intervals

N

1000ft

Fig. 163. Key to Fig .162

PHYSICAL PROPERTIES OF MINERALS

Mineral	Chemical Formula	Crystal System	Colour	Lustre
Orthoclase feldspar	$KAlSi_3O_8$	monoclinic	white, grey, flesh red, yellow, pink	vitreous
Plagioclase feldspar	albite: $NaAlSi_3O_8$ anorthite: $CaAl_2Si_3O_8$	triclinic	white, yellow, red, grey or black	vitreous
Quartz	SiO_2	hexagonal	colourless, white, rose, violet, smoky grey etc.	vitreous greasy
Muscovite	complex silicate of aluminium and potassium	monoclinic	transparent, grey, or light brown	pearly vitreous
Biotite	complex silicate	monoclinic	black, dark brown or green	pearly vitreous
Pyroxenes (e.g. augite)	complex silicates	monoclinic or orthorhombic	green, brown or black	vitreous, resinous dull
Amphiboles (e.g. hornblende)	complex silicates	monoclinic or orthorhombic	white, green, grey or brown	highly vitreous (may be
Calcite	$CaCO_3$	hexagonal	colourless, white, yellow etc.	vitreous dull
Dolomite	$CaCO_3.MgCO_3$	hexagonal	white, often tinged with other colours	vitreous, pearly o
Fluorite	CaF_2	cubic	colourless, white, blue etc.	vitreous
Aragonite	$CaCO_3$	orthorhombic	white, grey, green or violet	vitreous

~~E~~SCRIBED IN CHAPTER TWO

~~har~~dness	Specific Gravity	Diaphaneity	Cleavage	Fracture	Streak
6	2·5–2·6	transparent to translucent	two at right-angles	conchoidal	white, uncoloured
6	2·6–2·8	transparent to translucent	two good, clearly at right-angles; two others poor	uneven	white, uncoloured
7	2·65	transparent to opaque	none	conchoidal to uneven	white or pale
~~2~~–2½	2·8–3·1	transparent to opaque	perfect basal	—	—
~~2~~–2½	2·8–3·1	transparent to opaque	perfect basal	—	—
~~5~~–6	3·2–3·6	opaque	two almost at right-angles	uneven or brittle	white to greyish green
~~5~~–6	2·9–3·3	opaque	two inclined at 56° and 124°	—	pale or uncoloured
3	2·72	transparent to opaque	perfect rhombohedral in three oblique directions	conchoidal	white-grey
3½	2·8–2·9	opaque	perfect rhombohedral	conchoidal, brittle	white-grey
~~3~~–4	3–3·25	translucent	perfect octahedral	conchoidal	white
3½	2·9	transparent to translucent	one perfect, two imperfect	sub-conchoidal, brittle	white

Mineral	Chemical Formula	Crystal System	Colour	Lustre
Gypsum	$CaSO_4.2H_2O$	monoclinic	colourless, white, or stained	vitreous, pearly or silky
Anhydrite	$CaSO_4$	orthorhombic	colourless, white-grey	vitreous, pearly or greasy
Halite (rock salt)	$NaCl$	cubic	colourless, white, red, blue or violet	vitreous
Kaolinite	$Al_2O_3.2SiO_2.2H_2O$	triclinic	white, yellow or grey	earthy, greasy
Serpentine	$3MgO.2SiO_2.2H_2O$	monoclinic	dark, black-green	greasy to resinous
Chlorite	complex silicates	monoclinic	green	pearly
Bauxite	$Al_2O_3.2H_2O$	amorphous	white, yellow, brown, red	dull
Chalcopyrite	$CuFeS_2$	tetragonal	brass-yellow	metallic
Chalcocite	Cu_2S	orthorhombic	lead grey-black	metallic
Azurite	$2CuCO_3.Cu(OH)_2$	monoclinic	azure blue	vitreous
Malachite	$CuCO_3.Cu(OH)_2$	monoclinic	bright green	silky, vel dull
Native Gold	Au	cubic	pale to golden yellow	metallic
Galena	PbS	cubic	lead-grey	bright metallic
Cerussite	$PbCO_3$	orthorhombic	white	silky
Cinnabar	HgS	hexagonal	bright to brownish red	adamant to dull
Native silver	Ag	cubic	silver white, usually tarnished	metallic
Argentite	Ag_2S	cubic	blackish grey	metallic
Cassiterite	SnO_2	tetragonal	shades of green, brown, yellow, red or grey	adamant submeta and dull

dness	Specific Gravity	Diaphaneity	Cleavage	Fracture	Streak
2	2·3	transparent to opaque	one perfect, two imperfect	—	white
-3½	2·9	transparent to translucent	perfect in three directions	uneven to splintery	white
-2½	2·1–2·3	transparent to translucent	perfect cubic	conchoidal	white
-2½	2·2–2·6	opaque	perfect basal	crumbles	—
–4	2·5–2·8	translucent to opaque	poor	conchoidal to splintery	white
-2½	2·6–3	translucent to opaque	one perfect	—	greenish
–3	2·0–2·5	opaque	—	—	same as colour
–4	4·2	opaque	—	brittle	greenish black
–3	5·5–5·8	opaque	poor	conchoidal, brittle	greyish black
–4	3·8	transparent to opaque	—	conchoidal, brittle	light blue
–4	3·9–4·03	opaque	—	uneven	pale green
–3	1·9–3	opaque	—	hackly	yellow-white
2½	7·4–7·6	opaque	perfect cubic	sub-conchoidal	same as colour
3	6·4	transparent to translucent	—	conchoidal, brittle	white
-2½	8·1	subtransparent to opaque	perfect prismatic	sub-conchoidal to uneven	scarlet
–3	10·5	opaque	—	hackly	silver-white
-2½	7·3	opaque	—	sub-conchoidal	lead-grey, shining
–7	6·8–7·1	subtransparent to opaque	—	uneven, brittle	white to pale pale grey

Mineral	Chemical Formula	Crystal System	Colour	Lustre
Sphalerite	ZnS	cubic	yellow, brown-black	resinous adamantine
Haematite	Fe_2O_3	hexagonal	grey, red-brown or black	metallic earthy
Magnetite	Fe_3O_4	cubic	black	metallic submetal
Limonite	$Fe_2O_3.H_2O$	amorphous	yellow, brown or black	dull earthy
Pyrite	Fe_2S	cubic	brass-yellow	metallic
Pentlandite	$(Fe,Ni)S$	cubic	orange-yellow	metallic
Smaltite	$CoAs_2$	cubic	white	metallic
Chromite	$FeCr_2O_4$	cubic	black	submetal
Magnesite	$MgCO_3$	hexagonal	dull white	mainly vitreous
Uraninite	Complex UO_2	cubic	black or brownish black	pitchy, submetal
Carnotite	$K_2(UO_2)_2(VO_4)_2.3H_2O$	orthorhombic	brilliant canary yellow	earthy
Sulphur	S	orthorhombic	yellow	resinous greasy

dness	Specific Gravity	Diaphaneity	Cleavage	Fracture	Streak
–4	4·1	various	in six directions	conchoidal	light brown
6½	4·9–5·3	opaque, except in thin plates	poor	splintery to uneven	red
–6½	5·0–5·2	opaque	poor	conchoidal to uneven	black
5½	3·4–4·0	opaque	—	uneven	yellow-brown
6½	4·9–5·2	opaque	—	uneven	greenish or brownish black
–4	5·0	opaque	—	uneven	black
–6	6·4	opaque	moderate octahedral	uneven, brittle	greyish black
–6	6·4	opaque	—	uneven	brown
4	3	transparent to opaque	perfect rhombohedral	conchoidal	—
½	9·0–9·7	opaque	—	uneven to conchoidal	brownish black
soft	4·0	opaque	basal	uneven	pale yellow
2½	2·1	transparent to opaque	imperfect	conchoidal to uneven	white

WHERE AND HOW TO COLLECT ROCKS, MINERALS, AND FOSSILS

In rock, fossil and mineral collecting, as in most 'collecting' pursuits, the key to success lies in knowing where to look, what equipment to use and the most effective methods of collecting.

(1) Equipment

As a hobby, rock collecting is relatively inexpensive, because it requires a minimum of supplies and equipment. But even on the most casual basis there are certain items that must be acquired.

Hammer. The hammer is the basic tool in the collector's kit. Almost any moderate-sized hammer is satisfactory, but as collecting experience is gained you will probably want to get a geologist's hammer. These hammers, also called mineralogist's or prospector's picks, are of two kinds. One type has a square head on one end and a pick on the other; the second type is similar to a stonemason's or bricklayer's hammer, and has a chisel end instead of the pointed pick end. The square head of the hammer is useful in breaking or chipping harder rocks and in trimming large rock specimens. The chisel or pick end is good for digging, prying, and splitting softer rocks.

Collecting Bag. It will be necessary to have some type of bag in which to carry rocks and minerals, collecting equipment and other supplies. A knapsack, or similar canvas or leather bag, is most suitable.

Chisels. A pair of chisels are useful when specimens must be chipped out of the surrounding rock. Two sizes, preferably half-inch and one-inch, will usually suffice. A small sharp punch or awl is effective in removing smaller specimens from the softer rocks.

Wrapping Materials. Some specimens are more fragile than others, and these should be handled with special care. Always keep several sheets of newspaper in the collecting bag, and wrap each specimen individually as it is collected. Precautions taken in the field will usually prevent prized specimens from being broken or otherwise damaged. In addition to newspaper, it is wise to carry a supply of tissue paper and cotton wool in which to wrap the more fragile specimens. Small boxes or tins are also useful.

Map, Notebook and Pencil. It is most important to have some method of recording where your specimens were found. It is easy to forget where the material was collected, and one should never rely on memory. A small pocket-sized notebook is inexpensive and convenient to carry in the field. As regards maps, an Ordnance Survey map of the area with a one-mile-to-the-inch scale is the most useful. For more detailed work a six-inch map should be used.

Magnifying Glass. A magnifying glass or hand lens is useful when looking at small specimens. It will also prove helpful in examining larger rocks for small mineral inclusions. A ten-power magnification is satisfactory for most purposes, and several inexpensive models are available.

Paper, Polythene or Cloth Bags. Bags of varying sizes are useful containers for fossils and rock specimens. Locality data may be written directly on the bag, or on a label placed inside with the specimens. As an added precaution some collectors do both.

Other Useful Items. The items described above are those that are most needed, and constitute the basic collecting equipment. The more serious amateur may wish to include certain additional items which will place his collecting on a more professional basis. Some of these accessory items are:

A geologic map of the collecting area if one is available. Consult the Publication List of the Institute of Geological Sciences to see if a geologic report or map of the area has been published. Such maps can often be seen at libraries, museums, schools and colleges. They may also be purchased through a stationer.

A compass for more accurate location of collecting localities and for measuring the direction of dip; and a clinometer, a simple instrument for measuring the angle of dip.

Adhesive or masking tape. The locality information can be written on the tape and applied directly to the specimen.

Paper labels (about 3 × 5 inches). A properly filled-in label should identify each bag of material.

A pocket knife is useful in testing the hardness of rocks and minerals. A knife is also handy to dig fossils or mineral crystals out of soft rocks.

Acid (*it must be carefully handled*) is needed to test certain rocks to see if they are calcareous. Dilute hydrochloric acid (HCl) can be purchased at most chemists. It can be carried in one- or two-ounce dropper bottles which may also be purchased from chemists. A single drop of dilute acid is all that is necessary to test for the presence of lime. If the specimen is calcareous, it will effervesce. (Label the acid bottle POISON and keep it well out of the reach of children. If you should get acid on your clothes or skin, rinse it off with clean water as soon as possible.)

(2) Where to Look

Knowing where to look for specimens is a very important part of rock and mineral collecting. Rocks and minerals are all around us, and the problem is not so much knowing what to collect as what to leave out.

Try to take your rock samples from road and railway cuttings, rivers and gullies. Each bank, cliff, beach or excavation may yield interesting specimens of many different types. If the rock has undergone some degree of weathering, it will make the job of recovering the specimens a little easier. Weathering will also remove some of the loose surface material that might be covering the rock outcrop.

Mine dumps, rock pits and quarries are also good places to collect. Be sure to obtain permission to enter these localities (and all other private property), and work with extreme care, as such places may be dangerous. Be especially cautious of falling rocks and blasting.

Lava flows are good places to look for volcanic rocks. Many of the gas cavities (amygdales) are lined with quartz calcite, agate and other crystals.

Igneous rocks can be found in many parts of the country. Such intrusive bodies as granite, dykes or sills often yield many fine mineral crystals.

Look also in stream beds. Remember, however, that many of the rocks found here have been transported for great distances, and the specimens may not be native to the area in which they were collected.

Fossils are found under somewhat different conditions from rock and mineral specimens. We have seen that igneous and metamorphic rocks are not likely to be fossiliferous, and that most fossils are found in marine sedimentary rocks. These sediments were deposited under conditions that were favourable for organisms during life, and facilitated preservation after death. Limestones, shales and certain types of sandstone are typically deposited under such conditions.

Look particularly for areas where marine sediments have not been greatly disturbed by heat, pressure or other physical and chemical change. Try also to find places where the rocks have been exposed to weathering—this helps greatly to release the fossils from the enclosing rock material.

Quarries are good places to look, but—and this bears repeating—be sure to obtain permission before entering. Rock exposures in quarries are often fresh, but probably have still undergone some weathering. Bones and petrified wood are commonly found in sand and gravel pits associated with river terraces.

Pay particular attention to all cuttings, cliffs and beaches. Rocks exposed in this way are usually still in their original position and are fairly well weathered.

Stream beds are also good places to look; they are continually subjected to the process of erosion and new material is uncovered year after year.

If there are abandoned coal mines nearby, check the dumps of waste rocks around the mine shafts. A careful examination of the waste may reveal specimens of well-preserved plant fossils.

Occasionally bore-holes may have been made in the area in the search for oil or coal. The company responsible may let you have some of the material brought to the surface and may also supply you with information about what lies below.

(3) How to Collect

When a likely collecting spot has been picked out, examine the ground very carefully. See if there are any rock fragments which contain pieces of minerals or fossils, or are interesting in themselves.

If specimens have been freed by weathering, they can easily be picked up and placed in the bag. Often, however, it will be necessary to take the hammer and very carefully remove the surrounding rock. Smaller specimens may be more safely freed with the careful use of the chisel. Gently tap the chisel and gradually chip away the matrix—the rock that is holding the specimen. After most of the matrix has been removed, the specimen should be carefully wrapped and placed in the collecting bag.

Before leaving a collecting area, be sure to record its geographic location in the field notebook. Locate it on the map, then enter it in the notebook in such a manner that you, or another collector, could easily return to the same spot. If a county or topographic map is available, it is wise to mark the locality on the map. Next, write the geographic data on a label and drop it into the bag of material collected at that particular place. In addition, many collectors find it helpful to note the locality on the outside of each bag of rocks, minerals or fossils.

Material from separate localities should be kept in individual cloth or paper bags. Take every precaution to keep the labels with their respective specimens, and always remember that *a specimen without a locality is greatly reduced in value*. This is especially true if you have discovered a rare or valuable mineral.

Once again, you should *always* ask the owner's permission before entering or collecting on private property. You should respect his property, especially livestock and fences, and leave the area cleaner than you found it.

One of the best ways to learn how to collect is to take a field trip with an organized group, such as a museum class or a rock and mineral club. Here you will be seeing and working with competent collectors who can acquaint you with the fundamentals of field collecting. Contact also provides an opportunity for collectors to exchange rocks, minerals, and fossils, and tips on collecting and preparation and good collecting sites.

SYNOPSIS OF PLANT AND ANIMAL KINGDOMS

The following synopsis will provide the reader with a brief review of the major taxonomic units of the various kinds of plants and animals, as well as some indication of the order of their development.

Kingdom Plantae (Plants)

Sub-kingdom Thallophyta—plants not forming an embryo.
    Division Algae—diatoms, algae, and seaweeds.
    Division Fungi—bacteria and fungi.
Sub-kingdom Embryophyta—plants forming an embryo.
    Division Bryophyta—mosses and liverworts.
    Division Tracheophyta—plants with vascular tissue.
      Subdivision Psilopsida—simple rootless plants.
      Subdivision Lycopsida—club mosses and scale trees.
      Subdivision Sphenopsida—horsetails and their relatives.
      Subdivision Pteropsida—ferns, cycads, conifers, and flowering plants.
        Class Filicineae—ferns.
        Class Gymnospermae—cone-bearing plants.
          Order Pteridospermales—seed ferns.
          Order Cycadeoidales (Bennettitales)—cycadeoids.
          Order Cycadales—cycads.
          Order Cordaitales—the early conifers.
          Order Ginkgoales—ginkgos or maidenhair trees.
          Order Coniferales—pines, junipers, and firs.
        Class Angiospermae—flowering plants and hardwoods.
          Subclass Dicotyledoneae—oaks, roses, maples.
          Subclass Monocotyledoneae—grasses, lilies, palms.

Kingdom Animalia (Animals)

Phylum Protozoa—foraminifers, radiolarians.
      Class Sarcodina—one-celled animals with pseudopodia.
        Order Foraminifera—foraminifers or 'forams'.
        Order Radiolaria—radiolarians.
Phylum Porifera—sponges.
Phylum Coelenterata—corals, jellyfishes, hydroids.
      Class Hydrozoa—hydroids.
      Class Scyphozoa—jellyfishes.
      Class Anthozoa—corals and sea anemones.
Phylum Platyhelminthes—flatworms.
Phylum Nemathelminthes—roundworms.
Phylum Trochelminthes—rotifers.
Phylum Bryozoa—'moss animals' or 'sea mats'.
Phylum Brachiopoda—'lamp shells' or brachiopods.
      Class Inarticulata—brachiopods with unhinged valves.
      Class Articulata—brachiopods with hinged valves.

Phylum Mollusca—molluscs: clams, snails, squids.
    Class Amphineura—chitons or 'sea-mice' or 'coat-of-mail' shells.
    Class Scaphopoda—'tusk-shells'.
    Class Pelecypoda—clams, mussels, oysters, and scallops.
    Class Gastropoda—snails, slugs, and conches.
    Class Cephalopoda—squids, octopuses, the pearly nautilus, and the
        extinct ammonoids.
        Subclass Nautiloidea—nautiloids.
        Subclass Ammonoidea—ammonites.
        Subclass Coleoidea (Dibranchia)—squids, octopuses, cuttlefish,
           and the extinct belemnoids.
        Order Belemnoidea (Belemnitida)—belemnites.
Phylum Annelida—earthworms, leeches.
Phylum Arthropoda—crabs, shrimps, insects, spiders, ostracods, and the
    extinct trilobites and eurypterids.
    Sub-phylum Trilobitomorpha—extinct trilobite-like arthropods.
        Class Trilobita—trilobites.
    Sub-phylum Chelicerata*—scorpions, spiders, mites, 'horseshoe' or
        'king crabs', and the extinct eurypterids.
        Class Merostomata—'king crabs' and eurypterids.
           Order Eurypterida—eurypterids.
        Class Arachnida—scorpions, spiders and ticks.
    Sub-phylum Crustacea—crayfish, crabs, lobsters.
        Class Ostracoda—ostracodes.
    Sub-phylum Insecta—insects.
Phylum Echinodermata—'sea lilies', sea-cucumbers, starfishes, sea-
    urchins.
    Sub-phylum Pelmatozoa—cystoids, blastoids, and crinoids.
        Class Cystoidea—cystoids.
        Class Blastoidea—blastoids or 'sea buds'.
        Class Crinoidea—'sea lilies' and 'feather stars'.
    Sub-phylum Eleutherozoa—sea-cucumbers, starfishes, sea-urchins.
        Class Stelleroidea—starfishes and 'brittle stars'.
           Subclass Asteroidea—starfishes.
           Subclass Ophiuroidea—'brittle stars' and 'serpent stars'.
        Class Echinoidea—sea-urchins, heart-urchins, and sand dollars.
        Class Holothuroidea—sea-cucumbers.
Phylum Chordata—graptolites, fish, amphibians, reptiles, birds, and
    mammals.
    Sub-phylum Hemichordata—chordates with pre-oral notochord.
        Class Graptolithina (Graptozoa)—extinct graptolites.
    Sub-phylum Vertebrata—vertebrates: animals with a vertebral column.
        Superclass Pisces—fishes.
        Class Agnatha—lampreys and hagfishes.
        Class Placodermi—placoderms.
        Class Chondrichthyes—sharks, rays, and skates.
        Class Osteichthyes—bony fishes: perch, catfish, trout, eel.
        Superclass Tetrapoda—amphibians, reptiles, birds, and mammals.
        Class Amphibia—salamanders, frogs, toads.
        Class Reptilia—lizards, snakes, turtles, crocodiles, and the extinct
           dinosaurs, plesiosaurs, ichthyosaurs, mosasaurs, and ptero-
           saurs.

---

* The subphyla of Arthropoda are considered to be classes by some
authorities.

Order Cotylosauria—cotylosaurs.
Order Chelonia—turtles and tortoises.
Order Pelycosauria—extinct fin- or sail-back reptiles.
Order Therapsida—therapsids, or theromorphs.
Order Ichthyosauria—ichthyosaurs.
Order Sauropterygia—extinct marine reptiles with paddle-like flippers.
    Suborder Plesiosauria—Plesiosaurs.
Order Squamata—lizards and snakes.
Order Thecodontia—thecondonts.
    Suborder Phytosauria—phytosaurs.
Order Crocodilia—crocodiles, alligators, and gavials.
Order Pterosauria—flying reptiles.
Order Saurischia—lizard-hipped dinosaurs.
    Suborder Theropoda—bipedal carnivorous dinosaurs.
    Suborder Sauropoda—quadrupedal, primarily herbivorous dinosaurs.
Order Ornithischia—bird-hipped dinosaurs.
    Suborder Ornithopoda—duck-billed dinosaurs.
    Suborder Stegosauria—plated dinosaurs.
    Suborder Ankylosauria—armoured dinosaurs.
    Suborder Ceratopsia—horned dinosaurs.
Class Aves—birds.
Class Mammalia—mammals: opossum, bats, rodents, dogs, whale, horse, man.
    Subclass Allotheria—multituberculates.
    Order Multituberculata—small primitive rodent-like mammals.
    Subclass Theria—most of the living mammals.
    Order Insectivora—insectivores.
    Order Primates—lemurs, monkeys, apes, man.
    Order Edentata—tree sloths, armadillos.
    Order Carnivora—flesh-eating mammals: dogs, cats, seals.
    Order Pantodonta—extinct pantodonts or amblypods.
    Order Dinocerata—extinct uintatheres.
    Order Proboscidea—elephants, extinct mastodonts, and woolly mammoths.
    Order Perissodactyla—horses, rhinoceroses, extinct titanotheres.
    Order Artiodactyla—pigs, deer, camels, extinct entelodonts.

GLOSSARY OF TERMS USED IN GEOLOGY

A

AA—Blocky basaltic lava flow.

ABRASION—The process of wearing away by friction.

ACICULAR—Needle-like.

ACIDIC ROCKS—General term used in referring to quartz-containing igneous rocks; for example, granite.

ADAMANTINE—Having lustre like that of a diamond.

AEOLIAN—Refers to material deposited by wind which has transported particles from elsewhere; for example, loess, sand dunes.

ALGAE—simple plants of the division Thallophyta.

ALLUVIAL FAN—Deposit formed where a stream emerges from a steep mountain valley upon open, relatively level, land.

ALLUVIUM—Sediment deposited by running water.

ALPINE GLACIER—A stream of ice occupying a depression in mountainous terrain and moving towards a lower level; also called mountain glacier or valley glacier.

ALTITUDE—Height above sea level.

AMBER—A hard, yellowish, translucent, fossilized plant resin.

AMMONITE—Ammonoid cephalopod with complexly wrinkled suture pattern; member of subclass Ammonoidea.

AMORPHOUS—Without definite molecular structure; not crystalline.

AMPHIBIAN—Animals living on both land and water; for example, frogs and salamanders.

AMYGDALE—Gas cavities or vesicles in igneous rocks which have become filled with secondary minerals.

ANATOMY—The structural make-up of an organism, or of its parts.

ANGULAR UNCONFORMITY—See UNCONFORMITY.

ANHYDRITE—Calcium sulphate ($CaSO_4$).

ANTERIOR—Front or fore.

ANTHRACITE—Hard, very pure coal.

ANTHROPOLOGY—The study of man, especially his physical nature and the ways in which he has become modified.

ANTICLINE—An upfold or arch in the rocks.

ANTICLINORIUM—A broad anticlinal dome with minor folds and faults.

APERTURE—The opening of shells, cells, etc.

APHANITIC—Refers to rocks of such fine texture that the crystals cannot be seen with the naked eye.

AQUIFER—A porous, water-bearing rock formation.

ARAGONITE—Calcium carbonate ($CaCO_3$) crystallizing in the orthorhombic system. In shells it is chalky and opaque; is less stable than calcite.

ARCHAEO—Prefix meaning ancient; from the Greek word *archaios* ('ancient').

ARCHAEOZOIC—The oldest known geological era; early Pre-Cambrian time.

AREAL—Pertaining to area (for example, areal geology—the geology of a given area).

ARENACEOUS—Of the texture or character of sand—grain size, $\frac{1}{16}$-2 mm.

ARÊTE—a sharp ridge separating two cirques or glacial valleys.

ARGILLACEOUS—Clayey—grain size less than $\frac{1}{16}$ mm.

ARTEFACTS—Implements or objects made by man.

ARTESIAN WELL—A well in which water is derived from an aquifer overlain by impervious strata.

ARTICULATED—Joined by interlocking processes, or by teeth and sockets.

ASTERISM—Starlike pattern displayed in certain minerals.

ASTEROID—One of the many small planets between the orbits of Jupiter and Mars; known also as planetoids.

ASYMMETRICAL—Lacking symmetry.

ASYMMETRICAL FOLD—A fold in which the limbs dip at different angles.

ATMOSPHERE—The air surrounding the earth.

ATTITUDE—The position of a segment of rock stratum with respect to a horizontal plane (see also dip and strike).

AXIS—One of the imaginary lines in a crystal.

AXIS OF FOLD—an imaginary line passing through the crest of an anticline or the trough of a syncline.

B

BARCHAN—A crescent-shaped sand dune.

BARRIER BEACH—a low sandy beach separated from the mainland by a marsh or lagoon.

BASALT—A fine-grained igneous rock.

BASE LEVEL—Level of the body of water into which a river flows.

BASIC ROCK—Igneous rock containing a low percentage of silica; for example, basalt.

BATHOLITH—A huge mass of intrusive igneous rock with no known floor, more than 40 miles in diameter.

BAUXITE—the chief ore of aluminium; hydrated aluminium oxide.

BEDDING PLANE—The surface of demarcation between two individual rock layers or strata.

BEDROCK—Solid unweathered rock underlying mantle rock.

BELEMNITE—An extinct cephalopod related to the present-day squid.

BILATERAL SYMMETRY—Pertaining to the two halves of a body as symmetrical mirror images of each other.

BINOMIAL NOMENCLATURE—System of scientific nomenclature requiring two names: the generic and trivial names.

BIOGENETIC LAW—Law stating that ontogeny recapitulates phylogeny; the development of the individual recapitulates, or portrays, the development of the race.

BIOTITE—A type of mica occurring as dark (usually black) crystals.

BIREFRINGENCE—A property of minerals (other than those of the isometric system) that results in separation of light into two rays; known also as double refraction.

BITUMINOUS COAL—Soft coal.

BLACK LIGHT—Light produced by ultraviolet radiation.

BLASTOID—Stalked echinoderm with a budlike calyx, usually consisting of 13 plates; member of class Blastoidea.

BLOCK DIAGRAM—A three-dimensional sketch combining the surface geology and the front and side structure of an area.

BLOCK MOUNTAINS—Mountains formed by faulting.

BLOW-OUT—A relatively small, basin-shaped depression caused by wind erosion.

BOULDER—A more-or-less rounded rock with a diameter in excess of ten inches.

BRACHIOPOD—Bivalved marine invertebrate; member of phylum Brachiopoda.

BRACKISH—A mixture of salt and fresh waters.

BRECCIA—Rock composed of angular or broken fragments.

BRITTLENESS—Tendency of a mineral to break easily.

BRYOZOAN—Small, colonial, aquatic animal, usually secreting a calcareous skeleton; member of phylum Bryozoa.

C

CAINO—Prefix meaning recent; from the Greek *kainos* ('new').

CAINOZOIC—The latest era of geologic time, following the Mesozoic Era and extending to the present.

CALCAREOUS—Composed of, or containing, calcium carbonate; limy.

CALCITE—Calcium carbonate ($CaCO_3$) crystallizing in the hexagonal system. In shells it is translucent and more stable than aragonite.

CALDERA—A large basin-like depression formed by the destruction of a volcanic cone.

CALYX—In corals, the bowl-shaped depression in the upper part of the skeleton; in stalked echinoderms, that part of the body which contains most of the soft parts.

CAMBRIAN—the first (oldest) period of the Palaeozoic Era.

CARBONACEOUS—Containing carbon.

CARBONATE—Rock or mineral composed of carbon, oxygen, and other elements.

CARBONIFEROUS—The fifth geologic period of the Palaeozoic Era as recognized in Europe; comprises the Mississippian and Pennsylvanian Periods of North America.

CARBONIZATION—The process of fossilization whereby organic remains are reduced to carbon or coal.

CARNIVORE—A flesh-eating animal.

CAST—The impression taken from a mould.

CEMENT—Material binding together particles of sedimentary rocks.

CEMENTATION—Deposition of mineral material between rock fragments.

-CENE—Suffix meaning recent; from the Greek *kainos* ('new').

CEPHALON—The head; in trilobites the anterior body segment forming the head.

CEPHALOPOD—Marine invertebrate with well-defined head and eyes and with tentacles around the mouth; member of class Cephalopoda, phylum Mollusca; includes squids, octopuses, pearly nautilus.

CERATITE—An ammonoid cephalopod with suture composed of rounded saddles and jagged lobes; member of subclass Ammonoidea.

CHALK—A soft white limestone.

CHEMICAL WEATHERING—Weathering which results in a change in chemical composition; known also as decomposition.

CHERT—A compact, cryptocrystalline, flintlike variety of silica.

CHITIN—A hornlike substance, found in the hard parts of all articulated animals, such as beetles and crabs.

CHITINOUS—Composed of chitin.

CHLORITE—A greenish mineral chemically related to the micas.

CHROMOSOME—In organic cells, the bodies of tissue which carry the genes or hereditary determiners. There is a pair of chromosomes in each somatic cell and in each zygote; there is one chromosome in each gamete.

CINDER CONE—A steep-sided cone composed primarily of ash and cinders and formed by volcanic action.

CIRQUE—A large, deep, amphitheatre-shaped depression at the head of a glacial valley.

CIRRI—Usually applied to certain types of appendages formed by the fusion of setae or cilia.

CLASS—Subdivision of a phylum; a unit of biological classification.

CLASTIC ROCK—Rock composed primarily of rock fragments transported to their place of deposition.

CLEAVAGE The tendency of certain minerals to split in particular directions, yielding relatively smooth plane surfaces.

COASTAL PLAIN—A level plain composed of gently dipping wave- or stream-deposited sediments, with its margin on the shore of a large body of water; commonly represents an exposed part of the recently elevated sea floor.

COELENTERATE—Invertebrate characterized by a hollow body cavity, radial symmetry, and stinging cells; members of phylum Coelenterata include jellyfishes, corals, sea anemones.

COL—A saddle-like gap across a ridge or between two peaks.

COLONIAL—In biology, refers to the way in which some invertebrates live in close association with, and are more or less dependent upon, each other; colonial corals, hydroids, etc.

COLUMNAL—One of the disc-shaped segments of a crinoid stalk.

COLUMNAR SECTION—A diagram, drawn to scale, that illustrates the character of the formations by means of geologic symbols, the thickness of strata, and their order of accumulation.

COLUMNAR STRUCTURE—Parallel rodlike structure in igneous rocks.

COMPACT—Closely packed.

COMPACTION—The process whereby loose sediments are consolidated into firm hard rocks.

COMPOSITE CONE—A volcanic cone composed of alternate layers of lava and cinders.

CONCENTRIC—Having a common centre, as circles; refers to shell markings that are parallel to shell margin.

CONCHOIDAL—A type of fracture having curved concavities or the approximate shape of one-half of a bivalve shell. Glass shows excellent conchoidal fracture.

CONCRETION—An aggregate of nodular or irregular masses in sedimentary rocks and usually formed around a central core, which is often a fossil.

CONGLOMERATE—The rock produced by consolidation of gravel; constituent rock and mineral fragments are usually varied in composition and size.

CONIFERS—Cone-bearing shrubs or trees.

CONNATE WATER—Water trapped in a sedimentary rock at the time of deposition.

CONODONT—Minute toothlike fossils found in certain Palaeozoic rocks. Their origin is not definitely known, but they may have been part of some type of extinct fish.

CONSOLIDATION—Process whereby sediment is changed into solid rock.

CONSTELLATION—A group of stars that form a recognizable pattern.

CONTACT—The surface which marks the junction of two bodies of rock.

CONTACT METAMORPHISM—Metamorphism brought about as a result of the intrusion or extrusion of igneous rock materials, and taking place in rocks at or near their contact with a body of igneous rock.

CONTINENTAL SHELF—The relatively shallow ocean floor bordering a continental land mass.

CONTINENTAL SLOPE—The relatively steep slope between the edge of a continental shelf and the ocean deeps.

CONTOUR INTERVAL—The difference in elevation between two consecutive contour lines.

CONTOUR LINE—Line drawn on a map to join points at the same height above sea level.

COPROLITE—The fossil excrement of animals.

COQUINA—A porous, coarse-grained limestone composed primarily of broken shell material.

CORAL—Bottom-dwelling marine invertebrate that secretes calcareous hard parts; member of class Anthozoa, phylum Coelenterata.

CORALLITE—The skeleton formed by an individual coral animal; may be solitary or form part of a colony.

CORALLUM—the skeleton of a coral colony.

CORE—The dense innermost zone of the earth.

CORRELATION—The process of demonstrating that certain strata are closely related to each other, or are stratigraphic equivalents.

CORUNDUM—Aluminium oxide; the second hardest mineral known.

CRATER—The funnel-shaped depression in the top of a volcanic cone.

CREEP—The slow downward movement of soils and surface rocks.

CRETACEOUS—Youngest division of the Mesozoic Era.

CREVASSE—A deep fissure or crack in a glacier.

CROSS-CUTTING RELATIONSHIPS, LAW OF—A rock is younger than any rock across which it cuts.

CRUST—The outer layer of the earth.

CRYPTOCRYSTALLINE—Composed of very fine or microscopic crystals.

CRYSTAL—Regular polyhedral form, bounded by plane surfaces, that is assumed by a mineral under suitable conditions. Crystals have definite external symmetry and internal structure.

CRYSTALLINE—Possessing definite internal structure; not amorphous.

CRYSTALLOGRAPHY—The study of crystal structure.

CRYSTAL SYMMETRY—The number, location and balanced arrangement of crystal faces in reference to the crystallographic axes or other crystallographic planes or directions.

CUBE—Crystal form in isometric system; has six faces, each of which is perpendicular to an axis.

CUBIC—Having the general shape of a cube; the isometric crystal system is often called the cubic system.

CYSTOID—An extinct stemmed echinoderm with calyx composed of numerous irregularly arranged plates; member of class Cystoidea.

D

DECOMPOSITION—Chemical decay or breakdown of a rock; see also Chemical Weathering.

DEFLATION—The removal of loose rock and soil particles by wind.

DEFORMATION OF ROCKS—Any change from the original form of the rocks, for example, by folding or faulting.

DELTA—A roughly triangular, level deposit formed where a river enters a body of standing water.

DENDRITE—A branching or treelike figure produced on or in a rock or mineral, usually formed by crystallization of an oxide of manganese.

DENDRITIC—Branching or treelike in form.

DENTITION—The system or arrangement of teeth peculiar to any given animal.

DEPOSIT—Anything laid down, as sediment from a stream.

DEPOSITION—The process of depositing or dropping rock material.

DESICCATION—The loss of water from sediments.

DETRITAL—Composed of mineral or rock fragments.

DETRITAL ROCK—Sedimentary rock derived from fragmental material of other rocks; sand, mud, gravel, etc.

DETRITUS—Rock fragment remaining from the disintegration of older rocks.

DEVONIAN—The fourth oldest period of the Palaeozoic Era; follows the Silurian, precedes the Carboniferous.

DIAGONAL—Two-angled.

DIAMOND—Crystalline carbon, the hardest-known substance.

DIAPHANEITY—Relative transparency; the diaphaneity of a mineral is described as transparent, translucent, opaque, etc.

DIASTEM—A minor break in sedimentary rocks representing only a short length of geologic time.

DIASTROPHISM—Movements within the rocky crust of the earth.

DIATOMITE—A siliceous deposit composed of the remains of microscopic plants called diatoms.

DINOSAUR—Any of a large group of extinct reptiles which lived only during the Mesozoic Era.

DIP—The angle of inclination which the bedding plane of rocks makes with a real or imaginary horizontal plane.

DISCONFORMITY—See UNCONFORMITY.

DISINTEGRATION—Breakdown of a rock by mechanical means, known also as Physical Weathering or Mechanical Weathering.

DISSECTED—Cut into hills and valleys by erosion.

DISTILLATION—In fossils, that process by which volatile organic matter is removed, leaving a carbon residue as evidence of their existence.

DISTURBANCE—Regional mountain-building event in earth history, commonly separating two periods.

DIVIDE—A ridge or high area separating adjoining drainage basins.

DOLOMITE—A mineral composed of calcium magnesium carbonate, $CaCO_3.MgCO_3$.

DOME—A folded structure in which the beds dip outward in all directions from a central area.

DOUBLE REFRACTION—See BIREFRINGENCE.

DRIFT—Material deposited by a glacier.

DRUMLIN—Rounded streamlined mounds of till.

DUCTILE—Capable of being drawn into wire.

DUNE—A hill or ridge of sand formed by the wind.

DYKE—A tabular rock body, usually igneous in origin, which cuts across the surrounding rock strata.

E

EARTHQUAKE—A shaking of the earth's crust caused by the fracture and movement of rocks or by volcanic shocks.

ECHINODERM—A marine invertebrate with a calcareous exoskeleton and usually exhibiting a fivefold radial symmetry; member of phylum

Echinodermata; includes cystoids, blastoids, crinoids, starfishes and sea-urchins.

ECHINOID—Bottom-dwelling, unattached marine invertebrate with exoskeleton of calcareous plates covered by movable spines; member of class Echinoidea; sea-urchins, heart-urchins, sand-dollars.

ECOLOGY—The study of the physical and biological relationships of organisms.

EFFERVESCENCE—The fizzing reaction caused when a carbonate mineral is treated with acid.

EMBRYO—An organism in the early stage of development before birth.

EMBRYOLOGY—That division of biology which deals with the formation and development of embryos.

EMBRYONIC—Referring to the earliest undeveloped stage of an animal after the egg stage.

END MORAINE—See TERMINAL MORAINE.

ENDOSKELETON—The internal supporting structure of an animal.

ENVIRONMENT—The surroundings, physical, chemical, and organic, of an organism.

EOCENE—A division of geologic time, estimated to be the time from 50 to 40 million years ago; one of the older divisions of the Cainozoic Era.

EPICENTRE—A point or line directly above the focus or point at which an earthquake originates.

EPOCH—A division of geologic time, subdivision of a period.

ERA—A division of geologic time; includes one or more periods.

ERATHEM—In stratigraphy, the rocks formed during an era of geologic time.

EROSION—The wearing away and removal of soil and rock fragments by wind, water or ice.

ERRATIC—A boulder, transported by glacial action, which differs from the bedrock on which it rests.

ESKERS—Winding ridges of stratified sand and gravel deposited by streams running through or under a glacier.

ESTUARY—A drowned river valley where tidal effects are evident.

EVAPORATION—The process whereby a liquid becomes a vapour.

EVAPORITE—A sediment derived by chemical precipitation as salt-saturated water evaporates.

EVOLUTION—A term applied to those methods or processes and to the sum of those processes whereby organisms change through successive generations.

EXFOLIATION—The splitting of scales or flakes from a rock surface as a result of weathering.

EXOSKELETON—An external skeleton, or hard covering for the protection of soft parts, particularly among invertebrates.

EXPOSURE—An unobscured outcrop of rock appearing at the surface: see OUTCROP.

EXTRUSIVE ROCK—Igneous rock that has been extruded or forced out on to the earth's surface.

F

FACE—The outer surface of a crystal.

FACIES—A rock type or sequence, e.g. basin facies.

FAULT—The displacement of rocks along a zone of fracture.

FAULT BRECCIA—Breccia formed along the plane of a fault.

FAULT LINE—The line along which a fault plane meets the earth's surface.

FAULT PLANE—A fracture in a rock along which movement has taken place.

FAULT SCARP—A small cliff formed at the surface along a fault line.

FAUNA—An assemblage of animals (living or fossil) existing in a given place at a given time.

FAUNAL SUCCESSION—Succession of life forms through geologic history which shows that the life of any one period is different from that of preceding and succeeding periods.

FELDSPAR—A group of closely related silicate minerals, including orthoclase, microcline, sanidine, plagioclase, labradorite and others.

FIBRE—Any fine, thin, threadlike feature.

FIBROUS—Consisting of fibres.

FILAMENT—A fine thread or fibre.

FILIFORM—Thread-shaped, very thin.

FISSILITY—the tendency of certain rocks to split readily along closely parallel planes.

FISSION—The act of splitting or dividing into two parts.

FIORD—A deep, narrow, steep-walled inlet of the sea; formed by the flooding of a glaciated valley.

FLINT—An amorphous siliceous rock, usually dark and dull.

FLOOD PLAIN—A low area bordering a river which is covered by water when the stream is in flood stage.

FLORA—An assemblage of plants (living or fossil) existing in a given place at a given time.

FLUORESCENCE—Luminescence of a mineral during exposure to invisible radiation (such as from ultra-violet or X-rays).

FLUVIAL DEPOSIT—Sediment deposited by streams.

FOCUS—The point within the earth where an earthquake originates.

FOLD—A flexure or bend produced when rocks were in a plastic condition.

FOLDED MOUNTAINS—Mountains formed from folding of the rocks.

FOLIATED—Made up of thin leaves, as in mica.

FOLIATION—The foliated (layered) structure in metamorphic rock.

FOOTWALL—The rock beneath the hanging wall in an inclined fault.

FORAMEN—In brachiopods, the opening in the pedicle valve near the beak where the pedicle extends through the shell.

FORAMINIFERA—One-celled, generally microscopic animals which play a part in the deposition of limestone; an order of class Sarcodina, phylum Protozoa.

FORMATION—A rock unit useful for mapping and distinguishing, primarily on the basis of lithologic characters.

FOSSIL—A remnant or trace of an organism buried by natural causes and preserved in the earth's crust. (See also GUIDE FOSSIL.)

FOSSILIFEROUS—Containing fossilized organic remains.

FRACTURE—The texture of a freshly broken surface other than a cleavage surface, described as conchoidal, even, splintery, etc.

FRAGMENTAL ROCK—Rock composed of pieces of minerals or pre-existing rocks cemented together.

FUMAROLE—Vents or holes in volcanic regions from which gases issue.

G

GABBRO—Coarse-grained igneous rock consisting primarily of plagioclase, orthoclase and pyroxene.

GALAXY—An astronomical system consisting of thousands of millions of stars; for example, the Milky Way.

GANGUE—The worthless minerals in an ore.

GARNETS—A group of complex silicates characteristic of certain metamorphic rocks; may form perfect red glassy crystals.

GASTROLITHS—Highly polished well-rounded pebbles found associated with certain reptilian fossils; 'stomach stones'.

GASTROPOD—A terrestrial or aquatic invertebrate, typically possessing a single-valved, calcareous, coiled shell; member of class Gastropoda, phylum Mollusca; snails and slugs.

GEANTICLINE—A broad upwarp of the earth's crust, covering hundreds of miles.

GEIGER COUNTER—An instrument used to detect radioactivity.

GEM—A cut and polished gemstone.

GEMOLOGY—The science dealing with the study of gemstones.

GEMSTONE—A mineral suitable for cutting into a gem; the term gemstones is frequently used collectively to include both cut-and-polished stones and rough stones.

GENE—The basic building unit of heredity; a hereditary determiner.

GENETIC—Pertaining to origin.

GENETICS—That division of biology which deals with heredity and variation among related organisms.

GENUS—A group of closely related species of organisms. The plural is *genera*.

GEOCHRONOLOGY—The study of time in relation to the history of the earth.

GEODE—A rounded or spherical rock cavity; commonly lined with crystals.

GEOLOGIC AGE—The age of an object as stated in terms of geologic time; for example, a Carboniferous fern, a Cretaceous dinosaur.

GEOLOGIC MAP—Map showing distribution of rock outcrops, structural features, mineral deposits.

GEOLOGIC RANGE—The known duration of an organism's existence throughout geologic time, for example, 'Cambrian to Recent' for brachiopods.

GEOLOGIC TIME SCALE—Tabular record of the divisions of earth history.

GEOMORPHOLOGY—That branch of geology dealing with the earth's form; a study of landscape development.

GEOSYNCLINE—A great downward flexure of the earth's crust, usually tens of miles wide and hundreds of miles long.

GEYSER—A hot spring which erupts periodically, throwing out steam and hot water.

GEYSERITE—Siliceous deposits formed around the openings of geysers and hot springs.

GLACIER—A slowly moving mass of recrystallized ice flowing forward as a result of gravitational attraction.

GLASS—In geology, a brittle non-crystalline rock that has cooled rapidly.

GLASSY TEXTURE—Dense, non-crystalline texture like that of obsidian and certain other volcanic rocks.

GLAUCONITE—A greenish mineral commonly formed in marine environments, and essentially a hydrated silicate of iron and potassium.

GNEISS—A coarse-grained metamorphic rock having segregations of granular and platy minerals that give it a more or less banded appearance without well-developed schistosity.

GONIATITE—An ammonoid cephalopod with suture composed of rounded saddles and angular lobes; member of subclass Ammonoidea.

GRABEN—a long narrow block that has dropped down between two faults.

GRADIENT—The slope of a stream bed, usually expressed in feet per mile.

GRANITE—A granular igneous rock composed mostly of quartz, feldspar, and commonly mica or hornblende.

GRANITIC TEXTURE—Coarse-grained texture; characteristic of intrusive rocks.

GRANULAR TEXTURE—Texture marked by interlocking grains of similar size, as in granite.

GRAPTOLITE—An extinct marine colonial organism with chitinous hard parts; believed to belong to subphylum Hemichordata of phylum Chordata.

GRAVITY FAULT—See NORMAL FAULT.

GREENSAND—Sand or sandstone containing glauconite.

GROUNDMASS—Glassy or crystalline background material for the phenocrysts (larger crystals) in a porphyritic rock.

GROUND WATER—Underground water within the zone of saturation in the lower part of the mantle rock, known also as subsurface water, underground water, and subterranean water.

GUIDE OR ZONE FOSSIL—A fossil which, because of its limited vertical but wide horizontal distribution, is of value as a guide or index to the age of the rocks in which it is found.

GYPSUM—Hydrated calcium sulphate ($CaSO_4.2H_2O$).

H

HABIT—The characteristic crystal form of a mineral.

HACKLY FRACTURE—Tendency of a mineral to break with jagged, irregular surfaces.

HADE—The angle a fault plane makes with the vertical.

HALITE—Common rock salt, sodium chloride (NaCl).

HAEMATITE—Iron oxide (Fe_2O_3).

HANGING VALLEY—A tributary valley which enters the main valley at a greater elevation than the main valley floor.

HANGING WALL—The rock above the footwall in an inclined fault.

HARDNESS—The resistance of a rock to being scratched.

HARDNESS SCALE—A standard scale used to determine the relative hardness of minerals.

HERBIVORE—A plant-eating animal.

HEXA—A prefix meaning six.

HEXAGONAL—Having six angles and six sides; a crystal system in which the crystal faces are referred to four intersecting axes; three of these axes are equal and in the same plane, and intersect at angles of 120°; the fourth axis is perpendicular to the other three and is either longer or shorter.

HINGE LINE—In brachiopods, the edge of the shell where the two valves articulate; in pelecypods, the dorsal margin of the valve which is in continual contact with the opposite valve.

HISTORICAL GEOLOGY—Study of the geologic history of the earth.

HOMOLOGOUS STRUCTURE—Structures or organs in different animals that have the same fundamental structure, but are used for different purposes.

HOOK—A curved spit.

HORNBLENDE—A rock-forming mineral belonging to the amphibole group.

HORST—A block that has been raised between two faults.

HOT SPRINGS—Springs that bring hot water to the surface.

HYDROLOGIC CYCLE—The continuous process whereby water evaporates from the sea, is precipitated on to the land, and eventually moves back to the sea.

HYDROSPHERE—All of the water on the earth's surface or in the open spaces below the surface.

I

ICE AGE—The Pleistocene Epoch of the Quaternary Period, Cainozoic Era; a time of great glaciation.

ICE CAP—A localized ice sheet.

ICE SHEET—A large moundlike mass of glacier ice spreading in several or all directions from a centre.

ICHTHYOSAUR—A porpoise-like marine reptile of Mesozoic time.

IGNEOUS ROCK—Rock which has solidified from lava or molten rock called magma.

INCANDESCENCE—The glowing of a hot substance.

INCLUSION—Rock or mineral fragment surrounded by other rock.

INDEX FOSSIL—Same as Guide Fossil.

INTERMITTENT STREAM—A stream whose bed is dry part of the time.

INTRUSION—Igneous rock which, while in a molten condition, has forced its way into other rock.

INTRUSIVE ROCK—Rock that has been pushed (usually in a molten state) among pre-existing rock strata, commonly along faults or fissures. Intrusive rocks do not reach the earth's surface but are commonly exposed at the surface by later erosion.

INVERTEBRATE—An animal without a backbone or spinal column.

ISOMETRIC—A crystal system in which the crystal forms are referred to three equal axes intersecting at right-angles to each other.

ISOSEISMAL—An imaginary line connecting points of equal earthquake wave intensity; known also as isoseismic lines.

ISOSEISMIC LINE—See ISOSEISMAL.

ISOSTASY—The state of general equilibrium within the earth's crust.

ISOTOPES—Elements having the same atomic number, but differing in atomic weights and some properties.

J

JOINT—A fracture in a rock along which there has been no displacement on opposite sides of the break.

JOLLY-KRAUS BALANCE—A spring balance used especially for determining specific gravity.

JURASSIC—The middle period of the Mesozoic Era.

K

KAME—A small cone-shaped mound of stratified sand and gravel deposited by a glacial stream.

KAOLIN—A white or nearly white clay resulting from the decomposition of rocks containing large amounts of feldspar.

KARST TOPOGRAPHY—Irregular topography marked by sink-holes, streamless valleys, caverns, and underground streams.

KETTLE—A basin-shaped depression in glacial drift, formed when buried blocks of glacial ice melt.

L

LACCOLITH—A large lens-shaped mass of igneous intrusive rock.

LACUSTRINE DEPOSITS—Deposits formed on the bottom of lakes.

LAMELLAR—Arranged in plates.

LANDSLIDE—The relatively rapid movement of large masses of rock and earth down the slope of a hill or mountain.

LAPIDARY—Concerned with stones; a gem cutter.

LAPILLI—Small rounded or irregular fragments of volcanic rock thrown out during an eruption.

LATERAL—Side, or to the side.

LATERAL MORAINE—An elongated ridge of till along the lateral margins of an alpine glacier; derived largely from superficial debris falling on the glacier from the valley walls.

LAVA—Molten rock upon the surface.

LAVA DOME—See SHIELD VOLCANO.

LAVA PLATEAU—See PLATEAU BASALTS.

LIAS—Alternative name for the Lower Jurassic.

LIGNITE—Soft brown coal.

LIMESTONE—A sedimentary rock composed mostly of calcium carbonate.

LIMONITE—Amorphous hydrated iron oxide.

LITHIFICATION—the process whereby sediments are changed into solid rock.

LITHOLOGY—The study and description of rocks based on the megascopic (with the naked eye) examination of samples; used also to refer to the texture and composition of any given rock sample.

LITHOSPHERE—The solid part of the earth.

LOAD—The amount of material that can be transported by an eroding agent (such as a stream, glacier, or the wind) at a given time.

LODE—An unusually thick ore vein or group of ore veins that can be mined as a unit.

LOESS—A deposit of wind-blown silt.

LONGITUDINAL—In a direction parallel to the length.

LUSTRE—The appearance of the freshly broken or unweathered surface of a mineral in reflected light.

M

MACROFOSSILS—Fossils whose average representatives are readily visible to the naked eye.

MAGMA—Molten rock material beneath the earth's surface and from which igneous rocks are formed.

MALLEABLE—Able to be pounded or flattened without breaking.

MANTLE—The thick dense part of the lithosphere beneath the earth's crust extending to a depth of about 1,800 miles below the surface.

MANTLE ROCK—Layer of loose soil, earth, or rock which covers bedrock.

MARBLE—Recrystallized carbonate rock which, prior to being metamorphosed, was limestone or dolomite.

MARINE—Of, or pertaining to, the sea.

MARSUPIALS—Primitive mammals with young in pouches, now confined to Australia and America; e.g. opossums, kangaroos.

MASSIVE—Mineral habit lacking crystal form or imitative shape.

MASS MOVEMENT—Surface movements of earth materials caused primarily by gravity.

MASS WASTING—See MASS MOVEMENT.

MATRIX—The material in which a specific mineral is embedded; also the rock to which one end of a crystal is attached.

MEANDERS—A series of wide, looping curves in the course of a well-developed river.

MECHANICAL WEATHERING—See DISINTEGRATION.

MEDIAL MORAINE—An elongated ridgelike body of till formed by the junction of two lateral moraines.

MEGAFOSSILS—See MACROFOSSILS.

MESO-—A prefix signifying middle.

MESOZOIC ERA—Consists of Triassic, Jurassic and Cretaceous periods.

METAMORPHIC ROCK—Rock formed from igneous or sedimentary rocks that have been subjected to great changes in temperature, pressure and chemical environment.

METAMORPHISM—Extensive change of rocks or minerals.

METEORIC WATER—Ground water derived primarily from precipitation.

METEOROLOGY—The science which deals with the atmosphere.

MICAS—A group of rock-forming silicates.

MICROFOSSILS—Fossils which are, typically, microscopic in size.

MILKY WAY—The galaxy to which the earth belongs.

MINERAL—A naturally occurring inorganic substance possessing definite chemical and physical properties.

MINERALOGY—The science concerned with the study of minerals, including their occurrence, composition, forms, properties, and structure.

MIOCENE—The fourth epoch of the Tertiary Period of the Cainozoic Era. It lasted about 12 million years, following the Oligocene, and preceding the Pliocene.

MISSISSIPPIAN—The American equivalent to the Lower Carboniverous.

MOHOROVICIC DISCONTINUITY—The zone of contact between the rocky crust and the mantle; known also as the Moho.

MOHS' SCALE—Scale developed for determining the relative hardness of minerals.

MONADNOCK—An isolated hill left as an erosional remnant standing above the surface of a peneplane.

MONOCLINIC—A crystal system in which the crystal faces are described in relation to three intersecting unequal axes, two of which are at right-angles and the third inclined.

MONOTREMES—Primitive egg-laying mammals now confined to Australia; e.g. the duck-billed platypus.

MOON—A celestial body that revolves around a planet; a satellite.

MORAINE—An accumulation of rock materials carried and deposited by a glacier.

MORPHOLOGY—The study of structure or form.

MOUNTAIN—Any part of the land that rises conspicuously above the surrounding terrain; usually steep-sided and possessing a relatively small summit area.

MOUNTAIN GLACIER—See ALPINE GLACIER.

MUDFLOW—The movement of a large mass of mud, rock and water down a valley or stream course.

MUD VOLCANO—Bubbling springs of mud, often brightly coloured; also called paint pots.

MULTICELLULAR—Composed of more than one cell.

MUSCOVITE—'White mica'; characterized by silvery white crystals.

MUTATION—An inherited change transmitted as a result of changes within the germ plasm.

N

NATURAL SELECTION—Survival of organisms by reason of their ability to adapt to their surroundings and to changing environmental conditions.

NEBULA—A blurred, hazy mass of gases or dust in space.

NÉVÉ—Granular snow and ice which later becomes glacial ice.

NODULE—Rounded lump of rock or mineral.

NONCONFORMITY—See UNCONFORMITY.

NORMAL FAULT—A fault in which the hanging wall has moved downward with respect to the footwall; known also as a gravity fault.

O

OBLATE—Flattened at the poles.

OBSIDIAN—Glassy volcanic rock.

OCEANOGRAPHY—The study of the sea and its characteristics.

OCTO-—A prefix meaning eight.

OFFSHORE BAR—A sand bar that is more or less parallel to the coastline.

OIL SHALE—A highly organic shale from which petroleum can be extracted.

OLIGOCENE—A division of geologic time, estimated to be the time from 40 to 28 million years ago; part of the Cainozoic Era.

OLIVINE—A dark green, rock-forming silicate mineral.

ONTOGENY—Life history or development of an individual organism.

OOLITIC—Term applied to rock consisting of small rounded particles called oolites.

OPAQUE—Not transmitting light.

OPERCULUM—The lid, or covering, closing the opening of certain shells.

ORDOVICIAN—The second period of the Palaeozoic Era.

ORE—A metalliferous mineral deposit.

ORGAN—A part of a plant or animal that functions as a unit; for example, the heart, stomach, eye.

ORGANIC—Pertaining to, or derived from life.

ORGANISM—Any living being.

OROGENY—The process by which mountain structures are developed.

ORTHOCLASE—A feldspar characteristic of acidic igneous rocks; potash feldspar.

ORTHORHOMBIC—A crystal system in which crystal faces are referred to three unequal axes intersecting at right-angles.

OUTCROP—Place where bedrock is exposed at the surface.

OUTWASH PLAIN—A broad plain formed of deposits laid down by streams of a melting glacier.

OX-BOW LAKE—A crescent-shaped lake formed by the isolation of a meander from the main part of the stream.

OXIDATION—The chemical union of oxygen with other substances.

P

PAHOEHOE—A type of solidified lava characterized by a smooth, ropy or billowy surface.

PAINT POT—See MUD VOLCANO.

PALAEOCENE—The first epoch of the Tertiary Period.

PALAEOECOLOGY—The study of the relationship between ancient organisms and their environment.

PALAEOGEOGRAPHY—The study of ancient geography.

PALAEONTOLOGY—The science that deals with the study of fossils.

PALAEOZOIC—That era of geologic time containing the Cambrian, Ordovician, Silurian, Devonian, Carboniferous and Permian Periods.

PARACONFORMITY—See UNCONFORMITY.

PEAT—An accumulation of dark-brown, partially decomposed, plant material; the first stage in coal formation.

PEDICLE OPENING (PEDICLE FORAMEN)—See FORAMEN.

PEGMATITE—A body of coarse-grained intrusive igneous rock, commonly lens- or dyke-shaped.

PELECYPOD—A bivalved aquatic invertebrate; member of class Pelecypoda, phylum Mollusca.

PENEPLAIN (PENEPLANE)—An extensive, low, nearly flat region produced by continued erosion.

PENNSYLVANIAN—The American equivalent to the Millstone Grit and Coal Measures of the Carboniferous Period.

PENTAMEROUS SYMMETRY—Symmetry arranged in a pattern of fives.

PERIDOTITE—Coarse-grained basic igneous rock consisting primarily of olivine and pyroxene.

PERIOD—The division of geologic time which is next in rank below an era, and next above an epoch.

PERMEABLE—Capable of transmitting fluids.

PERMIAN—The last period of the Palaeozoic Era.

PERMINERALIZATION—In some fossils, that process by which mineral matter has been added to the original shell material by precipitation in the interstices rather than by replacing the original shell material.

PETROGRAPHY—The descriptive study of rocks.

PETROLOGICAL MICROSCOPE—A microscope for studying thin sections of rocks.

PETROLEUM—Oil; a complex mixture of hydrocarbons occurring in the earth's crust.

PETROLOGY—The study, by all possible methods, of the natural history of rocks.

PHENOCRYST—A crystal that is considerably larger than the crystals of the surrounding material.

PHOSPHATIC—Containing or pertaining to phosphate minerals.

PHYLOGENY—The racial history of a group of organisms.

PHYLUM—One of the primary divisions of the animal or vegetable kingdoms.

PHYSICAL GEOLOGY—The study of earth materials, their composition and distribution, and the forces which affect them.

PHYSICAL WEATHERING—The process of breaking rock down by physical forces; known also as mechanical weathering or disintegration.

PHYSIOGRAPHY—A description of the natural features of the surface of the earth.

PIEDMONT GLACIER—A glacier formed by the union of several alpine glaciers at the foot of the mountains from which the alpine glaciers originated.

PITCH—The angle between the axis of a fold and the horizontal plane.

PLAGIOCLASE—A mixture of feldspars containing sodium and calcium; occurs in both acidic and basic igneous rocks.

PLAIN—An area of low relief and low elevation underlain by essentially horizontal strata.

PLANET—An astronomical body that revolves around the sun in a regular orbit.

PLANETOID—See ASTEROID.

PLATEAU—A relatively flat area of high elevation underlain by essentially horizontal strata.

PLATEAU BASALTS—Great sheets of basalt which have flowed out of fissures in the earth's crust; known also as lava plateaus or flood basalts.

PLAYA—The dried floors of intermittent lakes found in desert areas.

PLEISTOCENE—Earliest epoch of Quaternary Period of Cainozoic Era; follows Pliocene Epoch of Tertiary Period, precedes Recent Epoch of Quaternary.

PLEURAL—Referring to the side or ribs; in trilobites, refers to lateral portions of thorax and pygidium.

PLIOCENE—Latest epoch of Tertiary Period of Cainozoic Era; follows Miocene Epoch and precedes Pleistocene Epoch of Quaternary Period; lasted about 13 million years.

PLUTON—Igneous rock bodies that solidified from magma at depth.

PLUTONIC ROCK—See INTRUSIVE ROCK.

POLYGONAL—Having more than four angles.

POLYP—A many-tentacled aquatic coelenterate animal, typically cylindrical or cup-shaped, as in corals.

PORCELLANEOUS—Like porcelain.

POROSITY—The percentage of porous rock that consists of open space.

POROUS—Containing pores or void spaces.

PORPHYRY—An igneous rock containing conspicuous phenocrysts in a glassy or fine-grained groundmass.

POSTERIOR—Situated behind; to the rear.

POTHOLE—A rounded hole ground in the rock of a stream channel.

PRE-CAMBRIAN—That portion of geologic time preceding the Cambrian Period of the Palaeozoic Era; divided into Archaeozoic and Proterozoic Eras.

PROTISTA—The organic kingdom including the simplest of all one-celled organisms which possess various characteristics of both plants and animals: bacteria, algae, foraminifers, radiolarians.

PSEUDOFOSSILS—Objects of non-organic origin which resemble fossils; for example, dendrites, concretions.

PSEUDOMORPH—A mineral which has assumed the shape of the mineral it replaced.

PSEUDOPODIUM—Temporary extension of protoplasm in certain one-celled organisms; used for taking in food, locomotion, etc.

PTEROSAUR—A flying reptile of the Mesozoic Era.

PYRITE—A hard, brass-yellow mineral composed of iron sulphide: 'fool's gold' (FeS_2).

PYROCLASTIC—Fragmental rock formed of rock fragments thrown out of volcanoes; for example, cinders, volcanic bombs.

Q

QUARTZ—A silicate mineral (SiO_2); the hardest common mineral.

QUARTZITE—Metamorphic rock which, prior to metamorphism, was sandstone.

QUATERNARY—The youngest period of the Cainozoic Era; follows the Tertiary Period.

R

RADIAL SYMMETRY—See SYMMETRY.

RADIOACTIVITY—Spontaneous disintegration of atomic nucleus, with release of energy.

RECAPITULATION, LAW OF—See BIOGENETIC LAW

RECESSIONAL MORAINE—Ridges of till formed by the recession (or temporary delays in the advance) of glaciers.

RECRYSTALLIZATION—Growth of small grains into large ones.

RECUMBENT FOLD—A fold in which the axis of folding is more or less horizontal.

RED BEDS—Red sedimentary rocks.

REEF—A moundlike or ridgelike elevation of the sea bottom which almost reaches the surface of the water; composed primarily of organic material.

REJUVENATION—Changes which tend to increase the gradient of a stream.

RELIEF—The irregularity of a land surface; the difference in elevation between the highest and lowest points in an area.

REPLACEMENT—Type of fossilization whereby organic hard parts are removed by solution accompanied by almost simultaneous deposition of other substances in the resulting voids; mineralization.

REVERSE FAULT—See THRUST FAULT.

REVOLUTION—Major mountain-building movement in earth history, typically of greater magnitude than a disturbance.

RHAETIC—An epoch in the Mesozoic intermediate between the Triassic and Jurassic Periods.

RHYOLITE—A fine-grained extrusive or shallow intrusive igneous rock of approximately the same composition as granite.

RIFT VALLEY—See GRABEN.

RIKER MOUNT—Paper-bound, glass-covered box for displaying specimens.

RIPPLE MARKS—Wavelike corrugations produced in unconsolidated materials by wind or water.

ROCK—Any naturally formed mass of mineral matter forming an essential part of the earth's crust.

ROCK-FORMING MINERALS—Common minerals which compose large percentages of the rocks of the lithosphere.

ROCK GLACIER—Lobate tongues of rock debris slowly moving downhill in the manner of a glacier.

ROCK SALT—Common salt (NaCl) or halite.

ROCKSLIDE—The relatively rapid downward movement of recently detached rock material sliding along zones of separation.

ROCK-UNIT—Divisions of rocks based on definite physical and lithologic characteristics and not defined on the basis of geologic time; groups, formations, members.

ROSSI-FOREL SCALE—Scale used to rate earthquake intensities.

RUN-OFF—Water that flows off the surface of the land.

S

SALT (common)—Sodium chloride; also called halite or rock salt.

SALT PLUG—Vertical pipelike bodies of salt (or gypsum) formed by the upward flow of salt under pressure; it has pushed up through the surrounding sediments in order to attain its present position.

SAND—Small mineral grains, usually quartz.

SAND DUNE—A ridge or hill of sand deposited by the wind.

SANDSTONE—Sedimentary rock consisting of consolidated sand.

SATELLITE—See MOON.

SCHIST—A metamorphic rock that contains an abundance of oriented platy minerals; the rock can be split with relative ease parallel to the flat surfaces of the plates.

SCLERITE—Minute skeletal element of sea-cucumbers.

SCOLECONDONT—The chitinous, horny, or siliceous jaws of worms.

SEA CAVE—Cave formed as a result of erosion by sea waves.

SEA CLIFF—Cliff formed by marine erosion.

SEDIMENT—Material that has been deposited by settling from a transportation agent such as water or air.

SEDIMENTARY ROCK—Rocks formed by the accumulation of sediments.

SEDIMENTATION—Deposition of the rock particles (sediments) that make up sedimentary rock.

SEISMOGRAM—The record made by a seismograph.

SEISMOGRAPH—An instrument used to record earth tremors.

SEISMOLOGY—The scientific study of earthquakes and other earth tremors.

SEPTUM—A dividing wall or partition; in fusulinids a partition between chambers in the fusulinid shell; in corals one of the radiating calcareous plates located within the corallite; in cephalopods the transverse partitions between the chambers.

SERIES—The rocks formed during an epoch; the time-stratigraphic term next in rank below a system.

SHALE—Thinly layered sedimentary rock composed of consolidated mud, clay, or silt.

SHIELD—A large area of exposed Precambrian rock.

SHIELD CONE—See SHIELD VOLCANO.

SHIELD VOLCANO—A volcano composed almost exclusively of lava; known also as shield cone, lava dome, or volcanic shield.

SILICA—An oxide (SiO_2) of silicon.

SILICEOUS—Containing or pertaining to silica.

SILICIFICATION—The process of combining or impregnating with silica.

SILICIFIED—Replaced by or containing a large amount of quartz or silica.

SILL—Solidified magma forced between layers of sedimentary rock.

SILT—Fine muddy sediment consisting of particles intermediate in size between clay particles and sand grains.

SILURIAN—The third oldest period of the Palaeozoic Era, follows the Ordovician, precedes the Devonian.

SINK—See SINK-HOLE.

SINK-HOLE—A surface depression in the ground caused by collapse due to solution of the underlying rocks; known also as a sink.

SLATE—Finely layered, compact metamorphic rock which splits readily into sheets; formed by metamorphism of shale.

SLICKENSIDES—Polished and grooved surfaces that are the result of two rock masses sliding past each other as in faulting.

SLUMP—The relatively small-scale downward slipping of a mass of rock or soil.

SMELTING—The process whereby ore is reduced to metal.

SNOW LINE—The level above which snow exists throughout the year.

SOIL—Broken and decomposed rock and decayed organic matter.

SOLAR SYSTEM—The sun and the astronomical bodies that revolve about it.

SOLIFLUCTION—Slow mass movement of soil especially characteristic of arctic or subarctic regions.

SOLITARY—Living alone; not part of a colony.

SPECIES—One of the smaller natural divisions in classification.

SPECIFIC GRAVITY—The weight of a material divided by the weight of an equal volume of water; also called relative density; the most commonly used standard temperature for its measurement is 4°C.

SPECIFIC NAME—The name applied to a species, usually the second of two names applied to a fossil, as *sapiens* in *Homo sapiens*.

SPHEROIDAL—Pertaining to a spheroid (a distorted sphere).

SPICULE—A minute spike or dart, skeletal element in sponges and holothurians.

SPIRE—The coiled gastropod shell exclusive of the body whorl.

SPIT—A finger-shaped sand bar extending out from the shore.

SPRING—Place where ground water reaches the surface through a natural opening.

STACK—An isolated steep-sided column of rock left standing as waves erode a shoreline.

STALACTITE—An icicle-shaped deposit formed by evaporation of solutions dripping from the roof of a cavern.

STALAGMITE—Icicle-shaped mineral deposit formed by the evaporation of solutions on the floor of a cavern.

STAR—A highly heated mass of incandescent gases, such as the sun.

STEINKERN—An internal mould.

STOCK—A circular or oval-shaped igneous body which increases in size with depth; has no known floor and an exposed surface area of less than 40 square miles.

STOPING—One of the processes by which intrusive igneous rocks enter the country rock; the magma works its way upward as blocks of country rock break off and fall into the magma where they are incorporated into the molten mass.

STRATIFICATION—Bedding or layering in sedimentary rock.

STRATIGRAPHY—The branch of geology dealing with the definition and interpretation of stratified rocks; especially their lithology, sequence, distribution, and correlation.

STRATOVOLCANO—A volcano characterized by a cone composed of alternating layers of lava and pyroclastics; known also as composite cone.

STRATUM—A single bed or layer of sedimentary rock.

STREAK—The colour of a mineral when finely powdered; usually determined by rubbing the mineral against a piece of unglazed porcelain.

STREAK PLATE—A piece of unglazed porcelain or tile used to determine the streak of a mineral.

STREAM CAPTURE—See STREAM PIRACY.

STREAM PIRACY—The diversion of water from one stream channel into that of another; known also as capture.

STRIATE—Bearing striations.

STRIATIONS—Closely spaced fine parallel lines.

STRIKE—The direction of a real or imaginary line that is formed by the intersection of a bed or stratum with a horizontal plane; strike is perpendicular to the dip.

STRIKE-SLIP FAULT—A fault in which the movement is essentially in the direction of the fault's strike.

STRUCTURAL GEOLOGY—That branch of geology which deals with the study of the architecture of the earth (the rocks and their relationships to each other).

STRUCTURE—Physical features of a rock such as jointing, bedding, banding, cleavage.

SUBGLACIAL STREAM—Stream flowing in a tunnel beneath a glacier.

SUBLIMATION—The process whereby a substance passes from a solid to a gaseous state and again becomes a solid, without first becoming a liquid.

SUBSIDENCE—Sinking of the earth's crust.

SUBSURFACE WATER—See GROUND WATER.

SUBTERRANEAN WATER—See GROUND WATER.

SUPERPOSITION, LAW OF—In an undisturbed sequence of rocks younger beds will overlie older beds of rocks.

SUSPENSION—The manner in which a stream carries material between the surface and the bottom.

SUTURE—The line of junction between two parts; in crinoids the line of junction between two plates; in gastropods the line of junction of the whorls as seen on the exterior of the shell; in cephalopods the line of junction between a septum and the shell wall.

SYMMETRICAL FOLD—A fold in which the axial plane is essentially vertical, resulting in limbs which dip at similar angles.

SYMMETRY—The reversed repetition of parts with reference to an axis.

SYNCLINE A trough or downfold in the rocks.

SYSTEM—In stratigraphy: the rocks formed during a period; the time-stratigraphic term next in rank above a series. In mineralogy: one of six divisions into which crystals are classified according to their symmetry.

T

TABULAR—Having a flat, relatively large surface relative to thickness.

TALUS—A mass of rock debris which collects at the bottom of a steep hill or cliff.

TARN—A small mountain lake formed in a cirque after removal of the glacial ice.

TARNISH—Surface alteration which may be seen in metallic minerals.

TAXONOMY—That branch of science which deals with classifications, especially in relation to plants, animals, or fossils.

TECTONIC EARTHQUAKE—Earthquake produced from tectonic crustal movements such as faulting.

TECTONIC MOVEMENTS—Movements resulting in deformation of the earth's crust.

TENACITY—The resistance of minerals to breakage, described by such terms as malleable, ductile, sectile and brittle.

TERMINAL MORAINE—Moraine formed at the point marking the farthest advance of the glacier; known also as end moraine.

TERMINATION—The end of a crystal that is completely enclosed by crystal faces; the crystal end that is not attached to the matrix.

TERTIARY—The oldest period of the Cainozoic Era; follows the Cretaceous Period of the Mesozoic and precedes the Quaternary Period of the Cainozoic.

TEST—The protective covering of some invertebrate animals.

TETRAGONAL—One of the six crystal systems.

TEXTURE—Physical appearance of a rock as indicated by the size, shape, and arrangement of the materials comprising the rock.

THECA—A sheath or case; in coelenterates the bounding wall at or near the margin of the exoskeleton; in echinoderms the main body skeleton (or calyx) which houses the animal's soft parts; in graptolites, any cup or tube of the colony.

THORAX—In trilobites, the part of the body between the cephalon and pygidium.

THRUST FAULT—A fault in which the hanging wall has moved up relative to the footwall; known also as a reverse fault.

TILL—Unstratified glacial deposit.

TILLITE—Rock formed from the consolidation of till.

TIME-UNIT—A portion of continuous geologic time; for example, eras, periods, epochs, and ages.

TIME-ROCK UNIT—Same as Time-Stratigraphic Unit.

TIME-STRATIGRAPHIC UNIT—Term given to rock units with boundaries established by geologic time; strata deposited during definite portions of geologic time; for example, erathems, systems, series, stages.

TOMBOLO—A strip of land so deposited as to connect a small island with the mainland or with another island.

TOPOGRAPHIC MAP—A map showing the physical features of an area, especially the relief and contours of the land.

TOPOGRAPHY—The physical features or configuration of a land surface.

TRANSPORTATION—The process by which rock materials are carried.

TRAP—See TRAP-ROCK.

TRAP-ROCK—A general term for certain dark-coloured igneous rocks such as diabase and basalt; also called trap.

TRAVERTINE—A form of calcium carbonate ($CaCO_3$) deposited by ground or surface waters; subterranean formations such as stalactites and stalagmites are composed of travertine; such deposits may also occur around the mouths of certain springs. Known also as TUFA.

TRIASSIC—The oldest period of the Mesozoic era; follows the Permian Period of the Palaeozoic and precedes the Jurassic Period of the Mesozoic.

TRICLINIC—One of the six crystal systems.

TRIGONAL—Three-angled.

TRILOBITE—An extinct marine arthropod having a flattened segmented body covered by a hardened dorsal exoskeleton marked into three lobes.

TRIVIAL NAME—The Latin name added to a generic name to distinguish the species; same as Specific Name.

TSUNAMI—A giant seismic sea wave generated by submarine earthquakes or other disturbances on the sea floor, known also as 'tidal wave'.

TUFA—Porous calcareous deposits accumulating around springs; calcareous tufa is sometimes called travertine.

TUFF—Rock formed from the lithification of volcanic ash.

TURBIDITY CURRENTS—Strong currents produced by muds sliding down continental slopes.

TWIN CRYSTALS—Two or more crystals which have intergrown in a definite way.

TYPE LOCALITY—The geographic location at which a formation was first described and from which it was named; or from which the type specimen of a fossil species comes.

U

UNCONFORMITY—A break in sedimentation due to erosion; a place in the earth's crust where eroded bedrock is covered by younger sedimentary rocks.

Angular Unconformity—Unconformity where the beds below the unconformable beds had undergone structural deformation before the overlying beds were deposited.

Disconformity—Unconformity where beds above and below the unconformable contact are parallel.

Nonconformity—Unconformity formed by deposition of sedimentary rocks on rocks of igneous origin.

Paraconformity—Unconformity characterized by even contact of parallel beds.

UNDERGROUND WATER—See GROUND WATER.

UNICELLULAR—Composed of one cell.

UNIFORMITARIANISM—The doctrine of the present as the key to the past; implying that geologic history is best interpreted in the light of what is known about the present.

UNSTRATIFIED ROCKS—Rocks that are not stratified or layered.

V

VALLEY—A generally elongated depression of the land surface which commonly contains a stream.

VALLEY GLACIER—See ALPINE GLACIER.

VALLEY TRAIN—A gently sloping flood plain formed of sediment deposited by water flowing out from the foot of a valley glacier.

VALVE—The one or more pieces comprising the shell of animals.

VARIETY—In mineralogy, a subdivision of a mineral species.

VARVES—Paired layers of sediments (one coarse and one fine) deposited in a glacial lake in a period of one year.

VASCULAR—Pertaining to tubes or vessels for circulation of animal or plant fluids.

VEIN—A more-or-less sheetlike occurrence of ore which has great length and depth but relatively little thickness.

VENT—The pipe or vertical hole through which magma rises in the centre of a volcano.

VENTIFACT—Stone that has been smoothed or changed in shape by wind action.

VENTRAL—Pertaining to the abdomen; as opposed to dorsal, pertaining to the back.

VERTEBRATE—An animal having a backbone or spinal column.

VESICULAR ROCK—Rock characterized by numerous small cavities caused by gas expansion.

VESTIGIAL STRUCTURE—Structure that has been reduced in size or usefulness during the course of evolutionary change.

VITREOUS—Glassy.

VOLCANIC ASH—The finest rock particles emitted during volcanic eruptions.

VOLCANIC BLOCK—A solid, angular fragment thrown out of an erupting volcano.

VOLCANIC BOMB—A spindle- or tear-shaped mass of congealed magma blown out during a volcanic eruption.

VOLCANIC GLASS—Non-crystalline rock formed by rapid cooling of lava.

VOLCANIC NECK—Solidified rock material formed by the cooling and hardening of magma in the central vent of a volcano.

VOLCANIC SHIELD—See SHIELD VOLCANO.

VOLCANISM—Effects of molten rock and volcanoes or volcanic activity; also known as vulcanism.

VOLCANO—An opening or vent in the earth's crust through which volcanic materials are erupted; refers also to the land form developed by the accumulation of volcanic materials around the vent.

VULCANISM—See VOLCANISM.

W

WATER GAP—Valley or pass through a mountain ridge through which a stream flows.

WATER TABLE—The surface below which all empty spaces within the rock are filled with water.

WEATHERING—The natural physical and chemical breakdown of rocks under atmospheric conditions.

WHORL—A single turn or volution of a coiled shell.

WIND GAP—A water gap that has been abandoned by its stream.

Z

-ZOIC—Suffix meaning life (from the Greek *zoe*, 'life').

ZONE—Part of a formation characterized by a particular fossil.

ZOOECIUM—Tube or chamber occupied by an individual of a bryozoan colony; also called an autopore.

APPENDIX E

SUGGESTED FURTHER READING

Introductory Texts

Age, D. V., *Introducing Geology*. Faber & Faber: London, 1961.
Bradshaw, M. J., *A New Geology*. English Universities Press: London 1968.
Cox, Barry, *Prehistoric Animals*. Paul Hamlyn: London, 1969.
Day, Michael, *Fossil Man*. Paul Hamlyn: London, 1969.
Evans, I. O., *The Observer's Book of Geology*. Frederick Warne: London, 1957.
Haywood, Helen, *The Days of the Dinosaurs*. Odhams Books: London, 1964.
Matthews, W. H., *Exploring the World of Fossils*. Children's Press: Chicago, 1964.
Read, H. H., and Watson, J., *Beginning Geology*. Allen & Unwin: London, 1966.
Rhodes, H. T., Zim, H. S., and Shaffer, P. R., *Fossils*. Paul Hamlyn: London, 1965.
Trent, C., *Exploring the Rocks*. Phoenix House: London, 1957.
Zim, H. S., and Shaffer, P. R., *Rocks and Minerals*. Paul Hamlyn: London, 1965.

General

Adam, F. O., *The Birth and Development of the Geological Sciences* Constable: London, 1954.
Adams, W. M., *Earthquakes*. Heath: London, 1964.
Bennison, G. M., and Wright, A. E., *The Geological History of the British Isles*. Edward Arnold: London, 1969.
Carrington, R., *A Guide to Earth History*. Chatto & Windus: London, 1956.
Fletcher, G. L., and Wolfe, C. W., *Earth Science*. Heath: London, 1959.
Gass, I. G., *et al.*, *Understanding the Earth*. Artemis Press (Open University textbook): Horsham, 1972.
Himus, G. W., and Sweeting, G. S., *The Elements of Field Geology*. University Tutorial Press: London, 1965.
Holmes, A., *Principles of Physical Geology*. Nelson: London, 1959.
Johnson, G., *Earthquakes and Volcanoes*. Lindsey Drummond: London, 1938.
Kirkaldy, J. F., *General Principles of Geology*. Hutchinson: London, 1958.
Kummel, B., *History of the Earth*. Freeman (British Regional Geology Handbooks, HMSO): London, 1970.
Miller, T. G., *Geology and Scenery in Britain*. Batsford: London, 1953.
Stamp, L. Dudley, *Britain's Structure and Scenery*. Collins (New Naturalist Series and Fontana Books): London, 1962.

Tank, R. W., *Focus on Environmental Geology*. Oxford University Press: Oxford and London, 1973.
Tarzieff, H., *Volcanoes*. Prentice-Hall: London, 1962.
Trueman, A. E., *Geology and Scenery in England and Wales*. Penguin Books: Harmondsworth, Middlesex, 1963.
Wells, A. K., *Outline of Historical Geology*. Thomas Murby: London, 1960.
Complete Atlas of the British Isles. Readers' Digest: London, 1965.
The Earth. Time-Life International: New York, 1964.

Mapping

Bradshaw, M. J., and Jarman, E. A., *Geological Map Exercises*. English Universities Press: London, 1969.
Bradshaw, M. J., and Jarman, E. A., *Reading Geological Maps*. English Universities Press: London, 1969.

Mineralogy

Cox, K. G., Price, B. N., and Harte, B., *Crystals, Minerals and Rocks*. McGraw-Hill: New York, 1969.
Kerr, P. F., *Optical Mineralogy*, McGraw-Hill: London, 1959.
Rutley, F., and Read, H. H., *Elements of Mineralogy*. 24th edn. Thomas Murby: London, 1960.
Smith, H. G., and Wells, M. K., *Minerals and the Microscope*. Thomas Murby: London, 1957.

Palaeontology

Andrew, R. C., *All About Dinosaurs*. W. H. Allen: London, 1959.
Beerbower, J. R., *Search for the Past*. Prentice-Hall: New York, 1960.
Black, R. M., *The Elements of Palaeontology*. Cambridge University Press: London, 1970.
Casanova, R., *Fossil Collecting*, Faber & Faber: London, 1960.
Eastern, W. H., *Invertebrate Palaeontology*. Harper Brothers: New York, 1960.
Davies, A. M., *An Introduction to Palaeontology*. Thomas Murby: London, 1961.
Matthews, W. H., *Wonders of the Dinosaur World*. Dodd, Mead: New York, 1963.
Rhodes, F. H. T., *The Evolution of Life*, Penguin Books: Harmondsworth, Middlesex, 1963.
Swinnerton, H. H. *Fossils*. Collins: London, 1960.
Woods, H., *Palaeontology*. Cambridge University Press: London, 1963.

Petrology

Harker, A., *Petrology for Students*. Cambridge University Press: London, 1960.
Hatch, F. H., and Rastall, R. H., *The Petrology of the Sedimentary Rocks*. Allen & Unwin: London, 1965.
Hatch, F. H., Wells, A. K., and Wells, M. K., *The Petrology of the Igneous Rocks*. Thomas Murby: London, 1961.
Tyrrell, G. W., *The Principles of Petrology*. Methuen: London, 1926.

Allied Subjects

Hanauer, E. R., *Biology Made Simple*. W. H. Allen: London, 1973.
Hess, F. C., *Chemistry Made Simple*. W. H. Allen: London, 1972.
Lewis, J., *Anthropology Made Simple*. W. H. Allen: London, 1969.
Money, D. C., *Climate, Soils and Vegetation*. University Tutorial Press: London, 1965.
Morris, D., *The Naked Ape*. Corgi Books: London, 1969.
Morris, D., *The Human Zoo*. Corgi Books: London, 1971.
Skinner, B. J., *Earth Resources*. Prentice-Hall: New Jersey, 1970.

Handbooks published by the British Museum, London

Brothwell, D. R., *Digging up Bones*.
Cole, Sonia, *The Neolithic Revolution*.
Le Gros-Clark, Sir W., *Fossil Amphibians and Reptiles* (1962).
Oakley, K. P., *Man the Toolmaker*.
Oakley, K. P., and Muir-Wood, Helen, *The Succession of Life through Geological Time* (1962).
Swinton, W. E., *Dinosaurs*.
Swinton, W. E., *Fossil Amphibians and Reptiles* (1962).
Swinton, W. E., *Fossil Birds* (1950).
British Caenozoic Fossils (1963).
British Mesozoic Fossils (1962).
British Palaeozoic Fossils (1964).

Regional Handbooks published by Her Majesty's Stationery Office, London

Northern England (1953).
London and Thames Valley (1961).
Central England District (1947).
East Yorkshire and Lincolnshire (1948).
The Wealden District (1954).
The Welsh Borderland (1948).
South-West England (1948).
Hampshire Basin and Adjoining Areas (1961).
East Anglia and Adjoining Areas (1962).
South Wales (1948).
North Wales (1960).
The Pennines and Adjacent Areas (1945).
Bristol and Gloucester District (1948).
Grampian Highlands (1948).
Northern Highlands (1960).
South of Scotland (1948).
Midland Valley of Scotland (1948).
Tertiary Volcanic Districts (1960).

Open University Publications

Science Foundation Course Units 22–7.
Science. A Second Level Course: Geology.

Index